The
Motivated
Salon

Mansfield
Beauty
Schools

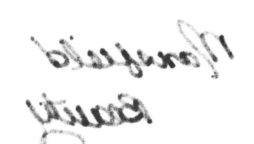

The Motivated Salon

by
Mark D. Foley

Milady Publishing
(a division of Delmar Publishers)
3 Columbia Circle, P.O. Box 12519
Albany, New York 12212-2519

NOTICE TO THE READER

Cover Design: Suzanne Nelson
Cover Photo–Shears: Photo courtesy Pivot Point International–Diamond 60 Shears

Milady Staff
Publisher: Gordon Miller
Acquisitions Editor: Marlene McHugh Pratt
Project Editors: NancyJean Downey and Annette Downs Danaher
Production Manager: Brian Yacur
Production and Art/Design Coordinator: Suzanne Nelson

COPYRIGHT © 1997
Milady Publishing
(a division of Delmar Publishers)
an International Thomson Publishing company

Printed in the United States of America
Printed and distributed simultaneously in Canada

For more information, contact:
SalonOvations
Milady Publishing
3 Columbia Circle , Box 12519
Albany, New York   12212-2519

1 2 3 4 5 6 7 8 9 10   XXX   01 00 99 98 97 96

Library of Congress Cataloging-in-Publication Data

Foley, Mark D.
     SalonOvations' the motivated salon / by Mark D. Foley.
        p.   cm.
     Includes index.
     ISBN: 1-56253-320-7
     1. Beauty shops—Management.   2. Beauty shops—Marketing.   I. Title.
TT965.F65   1996   96-15099
646.7'2'068–dc20   CIP

Contents

Acknowledgments

My career as a salon owner began in June 1988. I had just turned 27 years old. I owe a tremendous debt of gratitude to the gentleman who introduced me to the beauty business . . . the late Richard Vavra. His success as a multi-salon owner taught me that the salon business could be taken seriously and could yield rich rewards. He ranks as one of the most focused, determined, consistent, and good-hearted people I've ever had the privilege to know.

In September 1988 I met the late Neil Denholm, owner of Wild Rose Beauty. His guidance, encouragement, and friendship early in my career were instrumental in helping me achieve wonderful success as a salon professional. Also, it was Neil who gave me my start in facilitating seminars for salon colleagues when he invited me to address his clients in Edmonton, Alberta, Canada in September 1989. His level of professionalism, his total dedication to his clients, and his legendary ability to create rapid market expansion will always remain one of the greatest inspirations of my life.

There are so many among the living who have helped me and given me a forum for sharing strategies of salon success. Their number include distributors, manufacturers, trade show producers, magazine and book publishers, chain salon executives, and independent salon owners. I would like to list each and every one but we'd go on for pages. However, permit me to acknowledge Dale Schoeneman, Max Matteson, and Don and Flonnie Westbrook as folks who have had a particularly profound influence on my thinking and career. And, to all of you who have supported my efforts, I owe you a tremendous debt of heartfelt gratitude.

To the designers and salon owners and managers and estheticians and nail technicians and everyone who works in this glorious industry, thank you for the opportunity to be of service.

Catherine Frangie, Annette Downs Danaher, Marlene Pratt, and Gordon Miller of Milady Publishing and SalonOvations all have my sincere thanks. The opportunity and encouragement you gave me will be long remembered. Let it be known that the professionalism and commitment you all possess is one of the greatest resources we have in the salon profession. I salute you all!

About the Author

Mark D. Foley is considered by many to be the foremost motivational and business educator in the beauty business today as well as the most electrifying and dynamic seminar leader in the salon profession. His expertise in salon management, marketing, and customer service have won him clients representing a who's who of salon industry standouts.

Mark has written or been featured in articles by all the major salon industry trade magazines. He has personally trained more than 35,000 salon professionals internationally since 1990.

Mark's seminars "How to Double Your Haircolor Income . . . in 30 days or less!" and "How to Achieve Supernatural Salon Income" rank as all time best selling programs in the beauty business.

From a career as a highly successful owner of salons in Canada, Mark focuses his attention on writing, consulting, and speaking to communicate strategies for maximizing salon income and profits. An Eagle Scout and a graduate of Fordham University in New York, Mark makes his home in Calgary, Alberta.

Mark has been recognized with academic honors and awards from Harvard University, The University of Pennsylvania, New York University, UCLA, USC as well as the U.S. and Canadian governments. He's been the United States National Public Speaking Champion, the recipient of "The Golden Scroll Award" from The American Academy of Achievement and the recipient of "The George Washington Medal of Honor" from The Freedoms Foundation at Valley Forge.

Mark invites your questions, comments, and communications and offers to send you a current report "7 Key Ways to Dramatically Boost your Career and Business Growth," featuring income building strategies for salon professionals.

If you would like information on workshops, seminars, or other programs by Mark D. Foley, or to receive information on his library of salon income-building learning systems, please contact:

Mark D. Foley
1516 Locust Street
Denver, Colorado 80220

In Canada write:

Mark D. Foley
1235 Colgrove Avenue NE
Calgary, Alberta CANADA T2E 5C3

Or simply call toll-free **1-800-842-6241** from anywhere in North America.

Introduction

The best of times in cosmetology is right now! Colleagues who have worked in the profession for decades tell me regularly that we've never had it so good. Salon associates are making more money than ever before. The professionalism of our industry is at an all time high. Expansion is so dynamic right now that prospects for everybody are great!

In this environment, can you think of any reason to be a starving artist? Yet, as is true in all careers, we have people in the salon profession who are enjoying an abundant feast and others who are living on the crumbs.

If you're interested in climbing to new heights of career success then you're reading the right book. If you're a seasoned professional you'll learn and rediscover habits, attitudes, and philosophies that will add fresh vigor, more abundance, and greater satisfaction to your work life. If you're in the early years of your career, then you'll get on the fast track to success so you can situate yourself for higher income and rapid career advancement. Not only that, you'll arm yourself with information that will enable you to avoid the common mistakes that prematurely sidetrack thousands of young cosmetologists each year.

Your artistic ability with hair, skin, make-up, and nails is the cornerstone of your work. But only about 20% of your ultimate career success will have anything to do with your technical ability. The other 80% will depend on how you manage yourself and relate to others.

The primary purpose of this book is to reveal how you can enjoy financial prosperity and personal fulfillment in the beauty business. From this point forward I want you to remember that success is not something you pursue—rather, success is something you attract by the person you become. If we want to do better, we have to be better.

Much of what we're going to discuss will have more to do with working on yourself than working on your job. Ultimately, what I'm going to be sharing is how to get more out of life within the context of your career. And, in truth, your career is the ideal place to start because you devote a considerable amount of time, effort, and energy to it. However, keep in mind the good news that the person you become is not only going to influence your career—but also your entire life.

Improving your financial life is probably an area where you'd like to see immediate results. From wherever you're starting right now there are only three ways I know of that will dependably improve your salon income.

1. Serve more people.
2. Have the people you serve visit more consistently and frequently.
3. Have the people you serve spend more money with you.

As you can see, it's actually quite simple—there's no mystery here. The good news is that all these goals work hand in hand. By pursuing these objectives you're also pursuing the underlying cause of all prosperity—service! The more people you serve, the more frequently you serve them and the greater the measure of your service to them—the bigger an impact you're having on your fellow humans. And, the bigger an impact you have, the greater the value you bring to the marketplace. And, the greater the value you bring to the marketplace, the greater the measure of abundance and financial reward you enjoy.

How to execute these principles in your career is what we're going to discuss in some detail. And when we're done you'll be fully versed in the universal philosophy of business and personal success. You'll have hundreds of practical techniques you can use immediately to make a more meaningful contribution. You'll watch your own income grow as a consequence.

It's an exciting journey with many personal, professional, and financial rewards. When you think for a moment that the goal of our work is ultimately to help people look better and feel better about themselves, then it's easy to be enthusiastic about the value of our vocation.

PART I

EMBRACING THE VISION:

The Opportunity
for Success Has
Never Been Better

CHAPTER 1

The Opportunity Is Here and the Time Is Now

> **Good Luck is that point where preparation and opportunity meet!**

If you we're ever looking for a situation where you could be at the right place at the right time, I've got good news for you— you're there! The opportunity of a lifetime is at hand right now.

In the business of beauty, whether you're in cosmetology school or a 20-year salon industry veteran, your opportunity to enjoy personal happiness, financial prosperity, and professional fulfillment has never been better than it is right now! Your career in the image industry can provide you with everything that you can possibly want. Fame can be yours. Fortune can be yours. Let me assure you that this is not just idle optimistic chatter. This is as real as can be. You've got an unparalleled opportunity to enjoy the abundance of life beyond your wildest dreams. Believe it!

WHAT YOU WILL DISCOVER IN THIS CHAPTER

- The law of supply and demand is at work in the beauty business.

- The high and growing demand for salon services and products creates an opportunity for greater prosperity for salon professionals than ever before.

- Career prospects in the image industry present spectacular possibilities for both business and artistically minded colleagues.

- You're at the right place, doing the right thing, at the right time!

DEMOGRAPHICS DEMAND IT

We're at a unique time in human history. An overriding fact is the coming of age of the baby boom generation. This is a population that is fully committed to the ideal of looking good and feeling good about themselves. As they age, appearance services will become even more vital. And, because this group has created more wealth for themselves than any group before, they're in a position to afford the very best in beauty and grooming services.

When you're in a situation where you have a large population of people who are intensely interested in what you have to offer and can afford to pay for it, then you have the most essential ingredient for high levels of success. And that's just one of several factors pointing toward tremendous opportunity for all beauty industry professionals.

The prospects for the image business over the next 20 to 30 years are unbelievably positive. Image professionals will be in a position to make more money, have more clients, and experience more recognition and prestige than ever before. This remarkable situation all flows from the most basic law of economics.

THE LAW OF SUPPLY AND DEMAND

The Law of Supply and Demand in the Image Industry. In the salon business, we are the *supply*. A live person has to perform the services and operate the salons and the businesses that support them. The public creates the *demand* for our goods and services.

Right now we're in a situation where the demand for beauty services is so high and growing so rapidly that there's not a large enough supply of cosmetologists to provide quality services. The situation has become particularly acute in some geographic areas. Although this is unfortunate, it also creates a situation in which the opportunity for success is brighter than ever. Let's take a look at how this happened and discover the silver lining.

The Short Supply of Cosmetologists. In the recession of the early 1990s a tremendous number of cosmetology schools went out of business.

- First, government loans for students to attend cosmetology training became harder to get. The institutes themselves became responsible for making sure loans were repaid. Naturally, the academies became more selective in their student recruitment. Consequently, enrollments declined and some schools didn't survive.

- Second, the youth population in North America began to shrink by the 1990s. There simply weren't as many young people available to enter cosmetology school. The baby boom was over, and there was perhaps an overcapacity of schools relative to the youth population.

- Third, and perhaps most crucially, fewer people were interested in entering the cosmetology field. Someone, somewhere started a nasty rumor that cosmetology was not a high paying pursuit and lacked prestige and security. As a result of this misperception, cosmetology appeared less appealing as a career choice for many. Consequently, it was more challenging for schools to recruit students.

The Public Image. It isn't going to do us any good to deny that cosmetology has suffered from a low public image. It's a real issue. The issue has nothing to do with the validity of the assertion by some that cosmetology is a poor career choice. The opposite is true. Recent occupational forecasts have been very upbeat on the prospect of a career in cosmetology. Furthermore, cosmetology is one of the last areas of our economy where a person with little money or formal education, but with a lot of drive and ambition, can truly experience the full bounty of the American dream. I've seen it happen over and over again.

Rather, the poor public image (which is improving) of our field is significant because it brings with it the weight of shame, low self-esteem, and a lot of other negative rubbish. It can negatively affect our self-image as cosmetologists if we let it. That's the real problem! Any of this type of negative thinking is totally unproductive and we as individual professionals have to move beyond it if we're going to have any hope of making it to the top. This is an issue we will address throughout this book.

Combining these factors together has meant shortage. Recently the shortage has been so acute that thousands of entry level positions in the salon industry have gone unfilled year after year.

Furthermore, there has been a problem with keeping new graduates in the profession. Within 5 years of graduation fewer than half remain working in the industry. I've heard it said that only one-fifth make it past the 5-year mark. Whatever the numbers are, they're not good! Again, there are a variety of sad reasons for this which you'll find addressed throughout the book.

It seems that the tightening of entry requirements at the schools themselves will improve this figure, ensuring that only the most serious students will actually be admitted. However, it seems certain that as an industry we're not going to experience a boom in our workforce any time soon. It's clear that the supply of professionals to perform the services and run the salons is not expected to keep pace with the ongoing growth in demand for our services and products.

This current shortage of practitioners bodes well for those of us who continue to practice cosmetology in the years ahead.

This is so because there will be a continuous and substantial increase in the demand for our services.

Demand for Salon Services. The boom in demand for salon services and products has been going on for some time now and is picking up steam at a remarkable rate. From 1980 to 1990, revenues at salons in North America grew year after year after year. You've heard that the salon industry is "recession proof"? During the long and deep recession of the early 1990s our industry held firm and by the mid-1990s was picking up steam again.

The increased demand for salon services and products will continue. Two factors really substantiate this. First, our industry will grow in "market share." Second, the aging of the population will work to our benefit.

MARKET SHARE GROWTH

When I say we will grow in "market share" what I mean is that products and service that consumers often purchase outside of salons will more often be purchased in salons, for example, haircolor applications. Today, the large majority of haircolor applications are performed by consumers in their homes. However, that's changing. The growth in haircolor business at salons over the last number of years has been dramatic—some years reaching double-digit growth over the previous year. But we still have a long way to go and a lot of "market share" available for us to capture and bring into the salon.

Another example is salon retail products. Less than twenty cents of every dollar spent by consumers on shampoos, conditioners, and fixatives is spent in the salon. The rest is being spent in drug stores, supermarkets, and other outlets. Not long ago only a few pennies per dollar for these products was spent in the salons. As an industry, we've made great strides in capturing "market share." With the increased sophistication of manufacturers, improved consumer advertising, and enhanced salon merchandising and display efforts, our "market share" in this area will undoubtedly continue to expand.

Skin care, hair pieces, and nail products and services are other examples of areas that have enjoyed remarkable "market share" growth for salons over the last number of years.

LOW SUPPLY AND HIGH DEMAND

In the salon business we have a situation of a nongrowing supply of practitioners in the face of an absolute boom in demand, which creates a beauty business bonanza for a generation! Any economist will tell you that when you control a short supply in a high demand environment, you're operating in a particularly favorable economic predicament.

Well, we are the supply! In fact, economic theory would forecast an overall increase in salon prices and consequently in the income of all people working in salons. Ultimately, this will attract more professionals to our ranks.

The factors that led to the "identity crisis" and low image of a career in cosmetology have actually led to an environment allowing for greater prosperity than ever before. There's little doubt that as salon professionals continue to enjoy higher and higher levels of income, a greater measure of recognition and prestige will follow (Fig. 1-1). This, in turn, will attract more people to the cosmetology schools for training.

For right now, and for many years to come, cosmetology practitioners are in an unusually favorable economic situation. So if you had any question about it you can be 100% confident that you're in the right profession at the right time! Let's make sure each of us can reap a rich harvest during this season of bounty!

Your Career Possibilities are Tremendous. Cosmetology is growing at such a clip that the opportunities available to you as a professional are fantastic. Working in a salon and serving clients is the bed-rock of our business. But many cosmetologists pursue their careers outside the salon as well. There's the opportunity to work with product manufacturers and distributors, in education, on platform, to work on stage and screen productions, with photographers, or in salon management. You could become a celebrity. You can take your career anywhere you want it to go.

SUPPLY AND DEMAND IN THE SALON PROFESSION

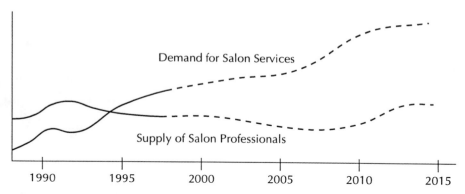

FIGURE 1-1 Growing demand for services with fewer professionals available to meet the demand leads to higher prices for services and higher income for salon professionals.

NEW MARKETING OPPORTUNITIES

How about new markets that the salon industry will capture? We're in the fashion business and things change fast. New service and product offerings, some of which are not even imagined today, will become fresh markets that salons will dominate.

I can remember a few years ago when hair extensions were all the rage. It was practically an unheard of service when it came on the scene. Soon salons across the country were doing millions of dollars worth of business in hair extensions. It was a new market that the salon industry dominated. This type of fashion phenomenon has been repeated many times and doubtless will continue into the future—further fueling revenues at salons.

THE "PRIME TIME" MARKET

Let's not forget that the aging of the population is another reason why business looks so bright for the coming years. The baby boom occurred from the end of World War II until the mid-1960s. It was the period of largest growth in our population and all these people have or will soon enter into their middle age. Research has discovered that the older people get (particularly women) the more time energy and funds they devote to beauty services. This only makes sense! We want to look as good as we can as long as we can.

The baby boomers have money to spend! That generation enjoyed a period of unprecedented economic growth and the highest incomes in history. Couple that with the fact that they're the beneficiaries of the largest transfer of inherited wealth in history and we have a population that not only wants, but can afford, the finest beauty services and products the world has to offer.

FINANCIAL OPPORTUNITY

We're now finding that meaningful incomes are being generated by salon professionals. As in the case with many professions, the first couple of years are often a "dues paying time." But we're seeing more salon professionals with only a couple of years experience earning over $30,000 annually. And, within 5 years we're seeing salon professionals reaching up to $50,000 annually. Superstar stylists earn in excess of $100,000 annually. I know this is true because I've met many high earners personally. I'm not talking about salon owners here. I'm talking about designers working in salons who are achieving this level of income.

I'm here to tell you right now that you can do it to!

Incomes of this level don't just happen automatically in any field. If we aspire to earnings of $50,000 a year or more we have to be outstanding at our profession, extremely dedicated, and fully prepared to contribute far more than average.

REACH FOR THE STARS

Of course, some stylists have achieved spectacular fortunes in the beauty business. Jheri Redding, Vidal Sassoon, Paul Mitchell, and Arnold Miller all became fabulously wealthy and all started their march to the top of the industry as stylists. Not only did they make a lot of money, but they had a tremendous impact on countless lives.

A whole host of cosmetologists have built wonderfully successful salons and salon chains through their spectacular business ability. Many independent salon owners have created remarkable enterprises and achieved millionaire status as a result.

It's just the beginning. A new generation of enterprising designers and business people will capture the imagination of the marketplace. New success stories are unfolding all the time. The environment is ripe for you to achieve your dream whatever it may be! (Fig. 1-2)

A HEALTHY DOSE OF PERSONAL SATISFACTION

We're fortunate because much of the happiness and enjoyment we experience in our careers flows from non-monetary sources. In addition to the opportunity to receive an outstanding income, we get many other payoffs as well.

Many of us have discovered the heartfelt joy we experience through our transformational service to others. Many of us know the self-esteem and self-respect that comes with being on the front line of the fashion industry. And don't forget the prestige and honor in your community that comes with a sparkling reputation for outstanding service and creative artistry. For many of us, these nonfinancial rewards are more important than the money we make.

SUMMARY

Make no mistake! The opportunity is here and the time is now. Here's what we discovered:

- The supply of salon professionals is lagging behind a rapidly growing public demand.

- Demand for salon services will continue to boom for the generation ahead.

FIGURE 1-2 Your dreams in cosmetology can come true. You hold the key in your hands.

- Tight supply in the face of heavy demand means happy days are here for those prepared to capitalize on their salon career.

- A life in the beauty business can take you anywhere you want to go—from big business to artistic excellence to the joys of contributing right in your own community.

- Your personal satisfaction from helping others is one of the most fulfilling rewards you'll enjoy!

The opportunity is there right now. Reading this book is one powerful way to prepare yourself to experience the full measure of benefit your career has in store for you. When you've got both preparation and opportunity coming together the lightening of good luck can't be far behind!

C H A P T E R 2

You Have an Unalienable Right to Pursue Success

> **66**
>
> **We hold these truths to be self-evident. That all men are created equal. That they are endowed by their creator with certain unalienable rights. That among these are life, liberty and the pursuit of happiness.**
>
> **99**

The immortal words Thomas Jefferson penned in the Declaration of Independence underscore the cornerstone of democratic life. What was Thomas Jefferson saying? He was saying that you have a right to freely pursue happiness. Furthermore, the right to pursue happiness is a God-given birthright dispensed equally, without discrimination, on each and every human soul! When you stop and think about it, Jefferson's words are quite profound.

Everybody has the right to pursue happiness. You don't have to be born into a certain family or go to a particular school or anything else. You have the right to pursue happiness simply by being a member of the human race. If you have the right to pursue happiness is there any good reason to be unhappy?

WHAT YOU WILL DISCOVER IN THIS CHAPTER

- You'll learn about the nature of happiness so it can grow in your life.

- You'll find out how to discover your purpose in life and understand how pursuing your life purpose is the key to real happiness.

- You'll discover the true nature of success and put aside some of the common myths that can slow you down.

- You'll crystallize your thinking on what you really desire in life.

- You'll begin to establish a cohesive vision for how you would like your life to unfold.

LET'S GET HAPPY!

What is happiness? That question can confuse many people. Let's take note of what some deep thinkers say. Success author Earl Nightengale said that "happiness is the progressive realization of a worthy goal or idea." Dr. William Sheldon states that "happiness is essentially a state of going somewhere wholeheartedly, one-directionally, without regret or reservations."

Real happiness has essentially three components. First, you have to comprehend and accept your purpose in life. That's the only thing worthy of spending your life on. Second, you have to set goals that act as milestones directing you toward your purpose. Third, you take action and make measurable progress in a reasonable time.

When you stop and think about it, this makes a lot of prac-

tical sense. You have to know who you are and what you're supposed to be doing in life. Then, you must make sure that you're
doing it! That's how happiness flows.

It seems that if you're not pursuing your heartfelt purpose
in life then you're swimming against the great stream of life.
That's when life becomes a struggle. That's when the joy of life
becomes elusive. That's when life becomes riddled with fear and
anger and resentment and bitterness and drudgery. That's no
way to live! When you're "going with the flow," progress is automatic and joyous.

CREATING HAPPINESS

You have to accept responsibility for your own happiness. It's up
to you to discover your purpose and move wholeheartedly toward
it. As a matter of fact, you must accept full responsibility for how
your own life unfolds from this moment forward. The acceptance
of personal responsibility is the cornerstone of successful living.
Taking responsibility for your personal progress adds a tremendous measure of freedom and adventure to life. You're vitalized!
Life has meaning! You're the captain of your own destiny and
that's highly exciting.

Where you are today is basically the sum total of all your
thoughts, attitudes, and actions accumulated from birth. It's
true, many outside influences early in life over which we had no
control set into motion patterns of behavior that continue. However, at some point in each person's life the individual must take
the reins and assume personal responsibility for designing the
rest of his or her life. Designing the future will involve undoing
some of the unmanageable habit patterns of the past. Certainly
by the time people reach the age of pursuing a career in cosmetology they can begin to take responsibility for their lives.

Reality is that you're responsible for creating your own
happiness. Here's some good news to keep in mind—your past
does not equal your future. For that matter, your present does
not equal your future. If you're not happy with your life the way
it is right now, accept the responsibility that you can start mov-

ing in a new direction at once. It's really as easy as that. The only thing that can prevent you from creating your own happiness is yourself.

CREATING YOUR VISION AND DESIGNING YOUR FUTURE

Life is like a hairstyle—with planning and performance it can be a masterpiece. You look at what's before you, you visualize where you want to go with it, you make a plan of action and, then, with one meticulous snip after another you wind up with a most pleasing design. A work of art—with happiness and joy for all involved. You have a feeling of satisfaction and accomplishment and all is right with the world. You're happy!

The same is true of life. With vision you can design your future, and with your day to day, moment by moment performance, you can make it happen. The whole process is fulfilling and creates a profound sense of happiness and joy in your life.

THE OPTION

Of course, the other option is to have no vision and make no plans. You wouldn't do it with a haircut or a manicure or a facial, but you'd be surprised how many people do it with their lives. There's an old saying that when you fail to plan you plan to fail.

Some people get into a rut of just taking life as it unfolds. Often they feel they are victims of circumstances and trapped by life. Sometimes they lose hope. Life has little or no meaning. Many of these folks have literally given up on life—they've relinquished their dreams—they've decided to let someone else make their plans for them.

In these situations, because there is no focus and no direction, there's little purpose. They go through the motions. Any sense of excellence and desire is lost. There's no joy. There's often a lot of worry about how things are going to turn out, but there's no initiative to create a better life.

In the civilized world, all of our basic needs are pretty much ensured. We'll eat. We'll have someplace to stay. We'll have something to wear. Many people figure this is comfortable enough so they "make do" and accept this common denominator. No real goals. No plans for their accomplishment. No burning desire. No activity above and beyond the essential. Not much of a future! What a sad state of affairs! Dare I say that the large percent of the population is in this quagmire.

IT STARTS WITH PURPOSE

Many times the reason folks are in this rut is that they have no good reason to change. "What's the use? Why bother?" That's a sure sign that these people have not found their purpose in life. If they became aware of their purpose in life then life would have meaning. The effort and energy required for advancement would make sense. Actually, a whole new spirit of life would be breathed into their activities.

I believe that we all have a purpose in life. This is a concept that has been repeated by all the great spiritual leaders, thinkers, and philosophers through the ages. If all the great minds and souls conclude that life has a purpose then the idea must have some merit, don't you think? People who discover their purpose in life seem to get much more accomplished because they have a reason to act like they mean it.

HOW DO YOU FIND YOUR PURPOSE?

You have a primary purpose in life and it's a good idea to get it down on paper so that there's no confusion about it. Let me guide you through some ideas to help you define your primary purpose.

1. *Happiness hints at purpose.* Your purpose in life is revealed by and is reflected by profound happiness. Think about the activities, people, and situations that bring genuine happiness to your heart. Magnify those experiences and notice the

common threads. Many of us in cosmetology get our greatest joy and satisfaction from our activities in the salon.

2. *Purpose is about service.* Another aspect of your purpose in life is that it will be outwardly focused. It will be about the happiness and satisfaction you derive from helping and serving others. Purpose in life is not about accumulation and possession. It's about the giving of ourselves for the benefit of others. You know the old saying "'tis better to give than to receive." The process of purposeful giving creates more inner happiness than anything else.

This is one of the reasons why I'm so excited about cosmetology. It's provides ample opportunity to make a profound and meaningful difference for other people—and, not only for our clients, but for our employees and coworkers as well. It's a real "people business" with a powerful people connection. It's so positive and potentially healing to the spirit and self-esteem of all with whom we connect.

3. *Goals are trail markers.* Your goals are not your purpose. Goals are milestones, trail markers, if you will, that we use to keep us on the path of our purpose. They're important because they help us direct our activities and gauge our progress. But they are not our purpose. We'll go into some depth on setting and achieving goals a little later.

4. *Your purpose can evolve and change.* Your purpose may change with the seasons of your life. You're life will always have purpose, but it's likely to undergo change and evolution as your life unfolds.

Your purpose in life does not necessarily revolve around your career. Your work may be a means for bringing your purpose about. For example, I know a hair designer from Vietnam who works for the First Place Chain in Birmingham, Alabama. She is dedicated to making life better for her children. Her purpose is clear. Her career in the salon, where she makes an income of more than $100,000 annually, enables her to provide for that heartfelt purpose.

Discover Your Own Purpose. The first place to start is by listening to your inner voice. People the world over meditate and
contemplate to get in touch with the voice within us all. Quiet
your mind, draw in your physical senses, close the mental
activity, take a few deep breaths, and ask your inner self what
your purpose in life is. You know that it's outwardly focused. You
know that it is accompanied by a sense of happiness.

In your heart of hearts, you will know your purpose. If
there's something you've had a hunch about—if there's something you've felt drawn toward—if there's something that you've
had automatic and natural interest in—you're on the right path.

Now lets put it down on paper in one clear sentence. You'll
fill in the blanks of this sentence:

My purpose in life is to:

PURPOSE STATEMENT
My purpose in life is to
(1) _____
(2) _____
(3) _____

In the first blank you put an action verb. It could be a word
like manage, help, organize, teach, or finance. In the second
blank you use a noun to describe who you're going to be acting
with. It could be an individual, a group, a class, an organization,
or a type of person. In the third blank you describe what you
undertake to accomplish. You define what it is that you want to
co-create, what you what to achieve, what progress you want to
make.

Let me give you an example of how this works by giving you
my own stated purpose for this particular season in my life: *My
purpose in life is to inspire beauty business professionals to success.*

It's clear and to the point. I know what my job is. I know who
I'm supposed to be working with and I know what we're supposed
to accomplish.

Now, do it yourself.

MAKE COSMETOLOGY YOUR PURPOSE

Many of you will naturally discover that your work in cosmetology is the purpose in your life. That's really great! Some of you will have other purposes. Fantastic! If you're unsettled and can't seem to get a clear fix on your purpose let me strongly advise you to select something. You can change it later on. But to move ahead effectively in life you must start somewhere.

In fact, under these circumstances, I'd suggest that you choose activity in cosmetology as your purpose in life. It's a noble pursuit. You serve other people. You must derive some happiness from it or you wouldn't have become involved to start with. Absent another compelling purpose, just make cosmetology your purpose in life—for right now. Just do it.

PROCEED WITH CONFIDENCE

Armed now with a full awareness of your direction, you can proceed toward it at once. Give yourself permission to go for it wholeheartedly. Don't hold back. Don't second guess yourself. You'll deny yourself ongoing happiness as long as you deny yourself the pursuit of your purpose. You have a purpose. Your life matters. Full steam ahead!

Another reality is that every purpose also carries with it the means for its own accomplishment. Your purpose, if it's genuine, can be achieved. Mother Nature will provide whatever you need to transform your purpose into reality. There's an old saying that the world will step aside to let the person proceed who knows where he is going. The way will be made clear. You will get what you need to accomplish the task. It's a cosmic mystery how all this happens. If the great stream of life is communicating that you need to move in a certain direction, then go in that direction fearlessly. Move ahead on faith. Purpose demands resolve. Resolve demands faith.

BE YOUR OWN BEST FRIEND

There's only one thing that can stand in the way of you pursuing your purpose. Take a look in the mirror to see the culprit. No one can stop you from doing what you are supposed to do except yourself. I know first hand that we can experience a lot of crazy thinking that denies us pursuing our purpose. That crazy thinking is rooted in fear. Fear of pain. The pain of failure, rejection, loss, struggle, effort. The fear that we're not good enough. The fear that we're not worthy. The fear of the pain that's associated with growth and change. Fear kills more dreams than anything.

Life is a struggle. That's reality. But not changing, not growing, not discovering, and not pursuing your purpose also involves pain. That pain is not only in the present but also in the future. That pain is chronic. It becomes the pain of regret, of "should've, would've, could've," and "if only."

The pain of change and growth is real. But it lasts only a season. The pain of staying the same is also real. But it lasts a lifetime. Ultimately, it's about valuing yourself enough and reflecting the love you have as a child of creation with a purpose for being. When you develop the willingness to go through the pain that can be associated with change and growth you'll discover miraculous results. It's about what happens when you turn your attention away from the darkness and toward the light. When you look to hope and possibility, rather than fear, the sunshine provides all the nourishment necessary to grow stronger and healthier.

MY PERSONAL STORY

Fear kept me from pursuing my career in cosmetology for over 3 years. After I graduated from Fordham University in New York City, I was immediately introduced to the salon industry. I knew at once that this is where I belonged. But what would people think? I was afraid of what people would think. I remember going to my father and telling him that I'd just discovered the beauty business. I related that for me it was thrilling and exciting to think about a career in the salon profession.

I didn't receive any encouragement at all. I got strange looks and expressions of disappointment and ridicule. I was afraid of disappointing my father. And if I had this reaction from him, I was afraid of what other people would think. That fear acted like a ball and chain around my ankle. Here I was, standing on the sidelines, wanting to be involved, but denying myself permission to do what I knew I was supposed to be doing! It sound ridiculous, but my fear held me back for over 3 years until circumstances unfolded and I became involved in salon ownership suddenly and abruptly before my family could talk me out of it.

Until I finally gave myself permission to pursue my purpose, I fumbled around at various jobs without any clear direction. I longed to be in the salon but denied myself what I wanted most. I was controlled by fear. And as long as I focused on the dark possibilities, I was paralyzed from moving ahead. Thank God, circumstances unfolded that threw me into the salon business suddenly. Otherwise I'm sure I would have always regretted not pursuing the profession I wanted in my heart.

PURPOSE BORN FROM PAIN

Sometimes your purpose in life can be born out of struggle. You've overcome something and you feel obliged to carry the message to others that they too can overcome the same challenge. That's part of my purpose—to share with fellow beauty professionals that not only is it OK and acceptable to pursue a career in cosmetology, it is an honorable and prestigious and excellent pursuit because it provides the ongoing opportunity to make such a profound contribution to others. Our career in cosmetology is not something we should whisper about with underlying feelings of shame and inadequacy. It is something we should take profound pride and pleasure in expressing. For those of us who've needed to go through that personal evolution in thinking—there is a tremendous amount of personal growth and acceptance to be experienced.

One of the greatest satisfactions I get is inspiring people to allow themselves to be "turned on" about being cosmetologists and giving themselves permission to pursue this career "all systems go" rather than think about it a just a job until something better comes along. For a number of years now I've hosted my highly successful "Sail to Supernatural Salon Success" cruise. It's amazing how many times spouses have come along on the cruise who'd been thinking about a career in cosmetology so they could be actively involved in the family business. As a result of the uplifting experience of the cruise, they finally give themselves permission to do what they had wanted to do, yet had denied themselves for years. It's amazing the freedom and joy that overcome people when they finally give themselves permission to pursue their purpose after years of painful denial. The progress they make is so rapid that it takes your breath away!

We owe it to ourselves to pursue our purpose without reservation. We do not need to apologize to anyone for our dreamed of career in cosmetology. You've heard of low self-esteem—imagine low career-esteem! We can have none of it! There can be no dark, negative thinking that belittles or devalues ourselves or cosmetology as a career. There can be no self-demeaning background noise in our mind clamoring of unworthiness or embarrassment. You must let the sunshine of purpose, pride, and opportunity radiate from your thinking. When you give yourself permission without reservation, only then are you free to pursue the full measure of greatness that is yours.

PURPOSE GIVES YOUR LIFE MEANING

Purpose and meaning go hand in hand. Your life has meaning because it counts for something. Other people are benefited. As baseball legend Jackie Robinson said "A life isn't significant except for its impact on other lives." Make the world a better place because you're here.

When you have a reason to get up in the morning and a reason for putting in the extra work and a reason for performing above and beyond the call of duty, you'll see that you can make massive progress quickly. Under these circumstances the little challenges of life are flicked aside like pieces of dust. A feeling of destiny takes over. The power and energy to act is available. As Jim Rohn said "when the reasons are clear the price is easy."

YOU HAVE WHAT IT TAKES FOR GREATNESS

I remember seeing a bumper sticker recently that proclaimed "God Don't Make No Junk." You have inherent value by virtue of your membership in the human race. Because you have tremendous potential to touch the lives of thousands of people in your career, you have the potential for greatness. Greatness in human life flows directly from the positive impact we have on others.

What is our purpose here? To discuss ways to make more money in the beauty business? Yes. To be successful at achieving our personal and professional goals? Yes. But it's just so much more than that. Fame and fortune are consequences of greatness. It's easy to get confused by the trappings of success and think that they are greatness itself. Success is not something that you obtain; success is something you attract by the person you become.

MYTHS AND TRUTHS

Many myths and truths surround success. It's important to clarify our thinking and understanding so that our approach to our career and our life can be sensible and true. Ultimately, what we're about here is discovering and developing our innate greatness and sharing the very best we have with others. It is a noble cause and it's your right to know the fundamental truths of success.

Do you want success? Of course you do! But what is it? The idea of success can be confusing for us because so many mixed signals bombard us. Your parents may have one idea of success. To your teachers in school success could be defined one way, to your social friends and peers success could be something altogether different. You may receive one message in church and another from society as a whole. And then we have the media and advertisers constantly overloading us with images and messages telling us we have to buy into some idea or obtain a product to be successful. Don't discount the power of all that advertising pressure. It is designed to make us feel inadequate and unsuccessful because we don't own the object product or aren't like the "successful" people being depicted. Many "myths" surrounding success need to be smashed so that we can observe success realistically.

THE FIVE COMMON MYTHS OF SUCCESS

1. *The "property" myth of success.* We haven't really arrived if we're not driving some luxury automobile or dripping with diamonds. We get the message that all we have to do is go out and obtain these objects and happiness will be ours.

But here's the real news—people who have gone through the process of obtaining all the objects of desire will report that owning them did not turn out to be the secret of happiness and success. They may be symbols of status and achievement—but in our easy credit society even that's questionable. The bottom line is that owning and possessing objects is not the definition of success. Money does not equal success. There are a lot of very unhappy, unfulfilled people with money running around out there. Yes, money is nice, and rewards along the highway of success are nice, but true success is a lot more fulfilling.

2. *The "well-born" myth of success.* Imagine buying into the idea that you have to be born into a certain kind of family or a certain social rank to be successful, that you have to have a certain family tree and grow up in an exclusive neighborhood to be really accepted a person of consequence.

Here are the facts. You're not born into success. You don't have to worry about your family background as an obstacle to enjoying success. In fact, if your family did not enjoy a tremendous amount of success and prosperity that may be all the more motivation for you to pursue it with real purpose. There's a lot to be said about being "hungry" for success!

There's no need to be a member of the social elite. The whole concept of social elitism is fading fast. Especially in North America we live where fortunes can rise and fall quickly. What previous generations accomplished or failed to accomplish need not have tremendous impact on your own personal journey. You are your own person and have open opportunity to make your own mark in the world.

3. *The "formal education" myth.* It should be clear that formal education or even intelligence are not prerequisites for experiencing all the success the world has to offer. Your ability to effectively deal with other people, your attitude, and your own sense of motivation, ambition, and determination are considerably more important than your formal education and native intelligence.

In the success literature we read of the IQ or "intelligence quotient" and the MQ or the "motivation quotient." The MQ is far more important to success than the IQ. I'm familiar with research that demonstrates that often people whose performance was only average during their school years became the greatest successes in their career and business life and that the academic wizards had only average careers.

In the beauty profession, and many other industries as well, I can point to legions of people who never graduated high school and yet accomplished a tremendous amount in their careers. Formal education is wonderful. However, it is not essential for success in cosmetology so long as people read the books and attend the seminars necessary for expansion of their knowledge base.

4. *The "your past equals your future" myth.* Your past does not equal your future. You can change at any time. This is an ongoing theme of this book. No matter what your past has been, you can re-invent yourself any time you choose to. It doesn't matter if you quit school. It doesn't matter if you've been to jail. It doesn't matter if you've been the victim of crime or abuse. It doesn't matter if you've been addicted to drugs or alcohol. It doesn't matter if you've had failed marriages. It doesn't matter if you've lived on the streets. You can transform just about any past

misfortune into the bedrock of strength on which you build your foundation for success.

Many people have had rough lives. Some of us have found it necessary to drag ourselves or let someone else drag us through the muck. The past is the past. The only thing that will ever change about the past is how we look at it. Part of success is knowing what you don't want! For those of us who have experienced a little misfortune, perhaps the bounty of success will be all the more sweeter. One thing is for sure, our success will be evidence of the tremendous amount of growth and transformation we've experienced.

5. *The success requires "good luck" myth.* First of all, luck is that point where preparation and opportunity meet. The harder and smarter you work the luckier you get. That's because your hard work has made you better prepared and put more opportunities in your path. It's as simple as that!

Focus, determination, the willingness to take a risk, and the right mental attitude are the ingredients of success. There's an old saying that you can't hit a home run if you're not up to bat. To experience luck and good fortune you must keep yourself in play. That's how you get opportunities. Also, seeing opportunities requires real experience, a dash of creativity and daring, and the right mental attitude. Lightning doesn't just strike and deliver long-term success to people. They work for it, believe me.

THE EIGHT TRUE CHARACTERS OF SUCCESS

Now that we've debunked some of the most common myths, let's try to understand the genuine nature of success.

1. *Success is a very personal matter.* Just as no two people are exactly alike, so too, no two definitions of success are alike. For one designer success may mean working on high-powered fashion shoots in Manhattan. For another, success may mean running a family-oriented country salon where you know everyone in town. And you couldn't convince either of them otherwise!

2. *Success is honest.* Not only does success have to do with being real and honest with other people, it also has to do with being honest with yourself. It has to do with coming to grips with what you really want out of life and not what other people or advertisers tell you. Sometimes it takes a little trial and error to figure out what works for you and what you're comfortable with, but so long as you're honest with yourself during the process you're on the path to success. Character and honor are cornerstones of success.

3. *Success is a process.* Yes, it's true that there are milestones to gauge our progress. Reaching those milestones is certainly cause for celebration. But enduring success is an ongoing phenomenon. It's a way of life. It's an ongoing process of continually moving in the right direction in balanced equilibrium.

4. *Success feels good.* Not only does reaching the milestones of success feel good, but striving to move forward feels good. That good feeling we have in our heart that we're pursuing something meaningful and worthwhile is what success is all about. It's what creates a sense of joy and serenity and ultimately defines the quality and character of an individual life. Doing and being good and contributing something of merit to others and providing well for ourselves and our family makes us feel good. It's a state of mind.

5. *Career success means satisfying work.* It seems to be true that the happiest people are those who do for a living what they enjoy most. I've often heard people with the spirit of success comment that they're amazed that they get paid to do what they love so much. This is particularly true in the arts—singing, acting, hair design, make-up artistry, fashion—a lot of satisfaction flows from the work itself. There's a feeling of completion and accomplishment. The reality of positive feedback from satisfied patrons enhances the feeling of satisfaction. We're so fortunate in cosmetology that our work is totally about helping people look good and feel good. It's gratifying to work at a pursuit that's so positive.

6. *Success pursues a purpose.* It bears repeating that a sense of purpose and meaning contributes to the feeling of suc-

cess. When that purpose is matched with strong and intense desire you have an unbeatable combination. Early on in life the purpose we pursue is often the desire for possessions or accomplishments. That's fine and it has its season. As a person matures that "something worthwhile" becomes less and less tangible. Lives become dedicated to higher principles. Purposes and activities are valued that embody those ideals. It's a process of maturity.

7. *Success requires contribution to others.* The desire to help and serve others is an ingredient of genuine success. It could have to do with providing for the family. In the cosmetology field there's a remarkable opportunity to contribute most meaningfully to the happiness and well-being of others. The feeling that we're doing something that's worthwhile and has merit contributes to a sense of success not only in our own hearts, but also in the eyes of others. Helping others achieve their goals is a key to success. As a matter of fact, the more people feel you're acting with integrity to benefit them, the more their feeling of trust and admiration toward you grows. That's when your leadership is strongest and most valued. It's a remarkable paradox that it's in your own self-interest to put other people first without sacrificing your own integrity.

8. *Success grows.* There's an old saying that when you're green you grow and when you're ripe you rot. Successful people are always striving to do and be more. They're always trying to improve themselves and enhance their effectiveness. It's remarkable that the successful people are great lifelong learners. They never stop advancing. They realize that when the school days are over, their real education is just beginning. In my seminars, it's always remarkable to notice that it's the professionals who are succeeding and striving for more that take the time to attend. The colleagues who need the information most aren't there.

I've learned that it's always easiest to do business with the people who are the most successful—it's their openness and receptivity to new ideas that's made them successful in the first place. They're looking for any edge that will make them more effective and they take on the perspective that they're delighted to spend an entire day in a seminar just to get one great idea that they can put to use. They're always striving to expand their "comfort zone."

DREAM YOUR DREAMS

We all have dreams of how we would like to have our life turn out. Our dreams tell us a lot about ourselves and about what we value in life. Our dreams are great tools of discovery and help us set our sails for the future (Fig. 2-1).

What I'm going to ask you to do right now is to have some fun and imagine and record your greatest dreams and aspirations. In this proces, I want you to believe that all things are possible. I want you to put aside any limitations or confinements. I want you to suspend reality as you understand it—but still stay within the earthly plane, fair enough? Dream wonderful and exotic dreams. Think big! There's magic in thinking big. Don't hold yourself back!

Imagine with me for a moment that you could wave a magic wand and miraculously create the life of your dreams. Visualize in your minds' eye what that life would be like 5 years from now, 10 years from now, 20 years from now. If you could receive everything in life that you want, what would you have for yourself? Have a vision of the life you want. Jonathan Swift wrote that "vision is the art of seeing the invisible."

It's valuable to translate your foggy, broad brush-stoke dreams into a more concrete form. That's the beginning of the process of defining your goals. But don't make the mistake of stepping back even one inch from your ideal vision. Go for it all! This is just an exercise, so don't you dare start limiting yourself at this point.

The problem that most people have is not that they aim too high, but rather that they aim to low and hit it. And then when they hit it, they stop. Big mistake!! This is a pitfall that successful and goal-oriented people can fall headlong into. They accomplish a series of goals that they had set out for themselves and then

FIGURE 2-1   What does success mean to you? Using your talents in service to others is how to attract the success of your dreams.

they stop. It's so vitally important to feed ourselves new and stimulating goals. That's the way to get yourself out of park and into gear if you're idling in the comfort zone. We all need to continually stretch and expand our goal offerings.

BE CLEAR

Now, be very specific about what it is exactly that you want to accomplish or what it is that you want to obtain. Be extremely precise. Call it the principle of specificity. Know for sure exactly what it is that you desire. Instead of "I want to attend a major salon trade show" proclaim that " I want to attend the International Beauty Show in New York City." Instead of "I want to increase my income" come right out and say "I want to earn $50,000 a year." Instead of "I want to work at a top salon" say "I want to work at the Grand Salon." Instead of having a goal of attending an advanced haircolor training program, get specific and decide which program specifically you want to attend. If you don't know which one, then make it a goal to find out!

Remember the old poem:

I bargained with life for a penny, and life would pay no more.
No matter how I begged in the evening when I counted my scanty store.
For Life is just an employer. It gives you what you ask.
But once you have set the wages, you must bear the task. I know.
I once worked for a menial's hire. Only to learn, dismayed.
That any wage I had asked of life, life would have willingly paid.

Here's your opportunity to get all your aspirations in a written form. The process of writing them makes them more tangible and real. On the following exercise forms notice the three headings: The person I want to be; The things I want to do; and The circumstances and possessions I want to have. Now consider your aspirations in terms of professional desires and personal desires. Contemplate what these possibilities mean to you and go for it.

WHO DO YOU WANT TO BE?

Write down all the things that you would like to be. Consider the sort of personality, character, skills, knowledge, ability and style that you'd like to call your own.

- Who do you admire most?
- What is it about them that you cherish?
- What accomplishments would you like to make?
- What talents would you like to develop?
- What sort of reputation do you wish to cultivate?
- What status do you want to enjoy?
- What sort of a manager or coworker do you want to be?

```
              THE PERSON I WANT TO BE
    1.  _____
    2.  _____
    3.  _____
    4.  _____
    5.  _____
    6.  _____
    7.  _____
    8.  _____
    9.  _____
   10.  _____
```

WHAT WOULD YOU LIKE TO DO?

Write all the things that you would like to do.

- What places would you like to travel to?
- What professional activities would you like to participate in?
- What hobbies would you like to learn?
- What experiences do you want?
- How do you want to be spending your time?
- Professionally, where do you want to go?
- Who do you want to associate with?
- Who would you like to meet?
- Who would you like to learn from?

```
THE THINGS I WANT TO DO

 1. _____
 2. _____
 3. _____
 4. _____
 5. _____
 6. _____
 7. _____
 8. _____
 9. _____
10. _____
```

WHAT WOULD YOU LIKE TO HAVE?

Write all the possessions and circumstances that you would like to have.

- What sort of professional and personal relationships would you like to have?

- What sort of romantic and home life?

- What material possessions would you like to have?

- What kind of home, wardrobe, automobile, furnishings?

- What sort of work environment?

- What financial situation?

- What investments?

- What sort of health and physical shape do you want to experience?

- What sort of spiritual life do you crave?

<table>
<tr><td colspan="2">THE CIRCUMSTANCES AND POSSESSIONS
I WANT TO HAVE</td></tr>
<tr><td>1.</td><td></td></tr>
<tr><td>2.</td><td></td></tr>
<tr><td>3.</td><td></td></tr>
<tr><td>4.</td><td></td></tr>
<tr><td>5.</td><td></td></tr>
<tr><td>6.</td><td></td></tr>
<tr><td>7.</td><td></td></tr>
<tr><td>8.</td><td></td></tr>
<tr><td>9.</td><td></td></tr>
<tr><td>10.</td><td></td></tr>
</table>

NOW BELIEVE

Now, believe that you can have it all because the plain fact is—you can have it all. You may discover that as time goes by you don't really want everything that you've written down. That's perfectly normal. Our wants and desires evolve as our experiences change. However, the fact is that you can experience the fulfillment of any one of those aspirations and you can experience the fulfillment of each of them. Accept the reality that it is within the abundance of your environment to provide you with each and every one of those intentions that you've written down.

At this point, it's not up to us to figure our how it's going to happen or how it could happen, but, merely to accept and believe that it can happen. And why not? Another one of the great teachings of the ages is that if you believe it, it will come to pass. James Allen wrote "as a man thinketh in his heart, so is he." The Bible teaches that "according to your belief is it done unto you." Merely the power of believing that it can happen automatically means that grace can bring it about . . . It can happen!

MAKE YOUR VISION CLEAR AND TRUE

LINE UP YOUR DESIRES WITH YOUR VALUES

Now that you've defined your purpose and listed your heart's desires, you can begin to refine your vision to make sure it's cohesive, consistent, and truly reflective of what you really want.

The first thing that you'll want to do is make sure your desires are consistent with your values. Your values are what you believe in, the ideals that are important to you and the way you want to live your life. Examples of values include your feelings about business and how business is to be conducted, your beliefs about how people should be treated, your priorities in life, and how you want other people to think about you and how you want to think about yourself as a person. All these ideas point to what your values are.

Some of the values that so many of us have are:

Love	Family
Service	Country
Honesty	Reverence
Justice	Friendship
Fairness	Security
Happiness	Thrift
Accomplishment	Honor
Charity	Trustworthiness
Wealth	Cheerfulness
Respect	Health

You'll want to make sure your desires promote and are in alignment with those values you hold dearest (Fig. 2-2). If your goals and values are out of alignment, you'll experience conflict that can easily dissolve the inner joy of happiness and make you feel like an impostor, even if you appear to succeed on the outside.

Here we're even entering into the realm of morality and ethics. Your values serve as a framework of evaluation to help you determine the merit of individual desires in the overall scheme of things. It's not unusual for us to have some quirky desires that may be, on second thought, better left alone. The reason is because they're not in harmony with the higher values that we are committed to.

LINE UP YOUR DESIRES WITH YOUR STATED PURPOSE

Evaluate how your desires relate to your purpose in life. With a career in cosmetology as your focus, you'll probably have many desires surrounding that pursuit—aspirations like developing the finest salon in town, introducing your own product line, cre-

Figure 2-2   For harmony and happiness, make sure your goals are in line with the values you treasure most.

ating a major innovation in hair design, traveling to and attending spectacular functions, or studying with the best educators in the field. Those are desires that relate specifically to your purpose. Whatever your purpose is, make sure an ample percentage of your desires are directed toward that end.

Naturally, we're not one-dimensional human beings. We have multiple wants. Sometimes even frivolous desires can strike us as being exciting even though they have little or nothing to do with our life purpose. That's great! If one of your desires is to visit the Pyramids in Egypt, then go for it!

DESIRES MUST BE COMPATIBLE

It's just important to understand that our desires, if they are to be happily accomplished, can't oppose each other in a practical way. For example, you'd have a real hard time enjoying a wonderful clientele as you operate the best salon in town on one hand, and then on the other hand become a professional golfer and tour the country playing the LPGA circuit. There are only so many hours in a day and we have to decide what is important enough for us to want to do.

For highly motivated big thinking people, multiple desires that contain practical conflicts can be frustrating. The key here is to make a decision about which one is important enough to focus your energies on and put the others away for a later time. If your energies are scattered and you're trying to serve too many masters, then nothing will get done well. You deserve better than that.

When deciding which conflicting desire to pursue, I'd suggest giving the nod to the one that's most in harmony with your life purpose. If you go the other way, you could suffer a constant low-grade gnawing of regret. You don't deserve that either. Go with your life desires; just make sure they're consistent and in harmony with your purpose and values.

SUMMARY

You have a natural right to pursue happiness. No one can take that away from you except yourself. Here's what we learned:

- Happiness is achieved by activating your life purpose.

- Purpose is outwardly focused and gives your life real meaning.

- The notion of success can be confusing, and ultimately it's a very personal pursuit.

- You can think big and attract greatness to your life if you start by defining exactly what you want to accomplish.

- It's vital that you give yourself permission to go for it all fearlessly.

- Making sure that your purpose and desires are in harmony will further create the mindset necessary for success.

Life is an adventure. There's so much to learn and do and experience. And cosmetology can take you anywhere you want to go. The reason this is so is that cosmetology gives us the profound opportunity to contribute to others. The universal law of success states that the greater our giving, the more we set into motion the forces of happiness and abundance in our own lives. You have the God-given right to pursue happiness. Can you think of any good reason not to? Go ahead and play full out, and enjoy yourself at the same time!

CHAPTER 3

Transforming Your Desires into Goals and Your Time into Results

66

There's plenty of room at the top. It's lonely up there!

99

Already we've come a long way—yet we've only just started! Our opportunity in the beauty profession to unbelievably positive. We're clear that we have a right to pursue happiness and enjoy the full banquet that life makes available for us. We've discovered the power of purpose and have begun to define what we want out of our careers and out of life. The next step is to set some specific goals.

You want to transform your desires into reality. The process of setting and achieving goals helps to get that done. Your goals act as mileposts to keep you on track and directed wonderfully as your life purpose unfolds. Goals are a way to ensure we're making measurable progress in reasonable time in the right direction.

WHAT YOU WILL DISCOVER IN THIS CHAPTER

- You'll learn the fundamental principles of setting and achieving goals.

- You'll clearly define where you want to go, how you're going to get there, and when you plan to arrive.

- You'll receive some valuable tips on using goals to keep yourself focused and motivated.

- You'll obtain a full gambit of ideas and concepts to help you better manage your time and make progress faster.

FUNDAMENTAL LAWS OF GOAL ACHIEVEMENT

Start with the destination. In the last chapter you had the opportunity to assess what you want out of life. You looked into the future without any limitations at all. Review that list of what you'd like to do, have, and be. Refine that list with what you've discovered about values and purposes to make sure your vision is unified. Now you have a good idea of what you want to accomplish and where you want to go. That's where you start.

It's important to know where you are so you know where you're starting from. But, don't let your current circumstances lower your sights one little bit. Don't worry about where you're starting from. You can get there from here! Let me show you how.

ESTABLISH YOUR TIME FRAMES

Put a time horizon next to each desire so you have a feel for how long it could take to transform that desire into reality. Some desires you can act on at once, others within a week or month. Some may take a year to achieve. Long-term goals can be 2, 3,

5, or even 10 years down the road. Don't worry now about how you're going to accomplish some of those long-term goals. We'll deal with that shortly.

Two points bear mentioning here. First, there is no such thing as an unrealistic goal; however, sometimes we put them in unrealistic time frames. Over time, you can make breathtaking advances and transformations. So don't cross something off your list because you're concerned about whether or not it is realistic—a lot can happen in 5 or 10 years! When you use the techniques and strategies discussed in this book, you'll be amazed at the rapid and triumphant progress you'll be making.

Sometimes we're a bit optimistic or naive and put goals into unrealistic time horizons. So what? That's not a big deal. If the time horizon passes and the goal is not achieved then simply set a new deadline. Don't give up. People have to adjust their time frames all the time.

However, the second point is that you want to be rigorous in setting time horizons. Challenge yourself. Make yourself stretch. It's been said that work expands to fill the time we have to achieve it. Avoid that trap. This is not a dress rehearsal. We've only got one pass through each moment. Set your goals in a way that forces you to get on with things and move along with some urgency. That's where you build the real momentum and make the rapid advancements.

ESTABLISH YOUR PRIORITIES

Not all desires are of equal value. Some things, quite naturally, are more important. Know what they are. Go over your desire list and put an A, B, or C next to each one. The A list contains the desires that mean the most to you—the one's you'd like to have transformed into reality most—the ones that will make the biggest difference in your life—the ones that move you clearly toward your heartfelt purpose. Your B list is made up of those that are of keen interest but of more secondary value. Your C list is made up of the desires that would be nice to have occur, but just aren't priorities.

Now look, all your desires can come to pass. The reason why it's good to prioritize is to make sure you're directing your energies where they matter most. You want to make sure you major in the majors and not in the minors. You want to avoid frittering your valuable time away on minor things. Your life is too important for that. Here is a principle that will recur in your journey of success, the 80/20 rule. In any list, such as a list of desires, you'll get 80% of your value by completing 20% of the items. It's important to understand which items are in the 20% that count so you can focus yourself there (Fig. 3-1).

MAKE YOUR GOALS SPECIFIC AND MEASURABLE

Be exact about what it is that you want. Clarify your vision. The more definite you can be the better. "I work at a great salon" isn't nearly as focused as "I work at the La Coiffure Grande Salon on Main Street." It's especially easy to describe tangible goals in great detail. Where exactly do you want to live? What neighborhood or street? What specific piece of property? What are the furnishings like? The rugs? The lighting fixtures? The wall paper? The tile? Make it vivid in every detail.

I've seen some people who have a goal about an automobile literally cut a picture of it out of a magazine and paste in on their bathroom mirror and tell people "that's my car." Even though they're telling the truth a little in advance it's OK! Being very specific let's you get full use out of the mind as a goal-achieving mechanism.

Frame your goals so that they can be measured using objective criteria—how many, how much, what percentage, and so forth. That's the best way of making them concrete. You can then measure your progress over time. Now, I understand that not all goals can be defined this way—like falling in love and marrying the one of your dreams—but the fact is, many goals can be. It's worthwhile to make every effort to let the numbers do the talking to make sure you're proceeding toward your goals by making measurable progress in reasonable time.

Figure 3-1 Focus your energies on the 20% of activities that yield the big payoff rather than on the 80% of activities that yield little reward.

Let's say, for example, that trip to The Long Beach Beauty Expo is going to cost you $1000. That number is something you can visualize. You can plan to pay your Long Beach Expo Fund $10.00 from every haircolor you do. That means you have to do 100 haircolor applications to get your $1000. Now, your goal is specific and measurable and achieving it becomes more exciting. Setting aside those funds and attending to the travel details and other fun things you want to do while in Southern California now becomes tangible and real and is something you can contribute to each day. You want to break your desires down in a way that can be measured so you can gauge your progress!

KEEP YOUR GOALS PRIVATE

As we get serious about setting and accomplishing meaningful and challenging goals, we sometimes feel compeled to share them with others. Generally that's a serious mistake. The sad reality is that many people will discourage us and tell us that we're going to fail and waste our time. The reasons people do this are varied and complex. So though I won't analyze the motives, it's important for you to be alerted to this phenomenon.

Unfortunately, it's often the people closest to us who will take the wind out of our sails. With a close relative it sometimes takes no more than a slight facial expression or change in body posture to convey a message of doubt and uncertainty. Worthwhile goals have been abandoned over no more than a few discouraging words spoken by another. Avoid that possibility. Not only must you be careful about who you share your plans with, but you must be careful who you share your progress with as well. The same people who will discourage you from your goals may be apt to belittle and minimize your accomplishments. Keep yourself out of the line of fire.

Make your goals and their achievement your own personal and private adventure. If you're going to share your goals with anyone make sure it's someone who will be supportive and encouraging. Generally, that means someone who's already accomplished several goals themselves. Choose your confidants carefully. Make your confidant a mentor, someone you'd like to be like and who can actually help show the way to get where you want to go!

DETERMINE THE INFORMATION AND PEOPLE YOU'LL NEED

Practically every worthwhile goal is going to require you to learn something new. That's great! Never let the fact that you'll need to expand your knowledge base and professional contacts slow you down. Resolve to get the information you'll need to proceed successfully. You owe it to yourself to take full advantage of the available knowledge and resources.

It's been proven time and again that proper preparation prevents a poor performance. The old saying that knowledge is power is true. Although you don't want to run into a situation of "paralysis by analysis," always be sure to give yourself the advantage by reading a book or two and taking a seminar or two on the topic of interest before rocketing headlong into a project. "On the project" education can often result in costly mistakes and unnecessary delays when problems could have been avoided by learning from the experience of others in advance. Don't neglect information gathering as part of your goal achievement process.

CREATING YOUR FORMAL GOAL LIST

REORGANIZE AND REWRITE YOUR GOALS

Now let's take the time to put down on paper your formal goal list. It's been mentioned time and again that your goals must be established clearly in writing. Up in your head is not good enough. You want them written down so you can refer to them often. They're more concrete when they're written. They're more real.

Notice on the next page the worksheet with the three columns with the headings: "The Kind of Person I Am," "The Activities I Do," "The Wealth I Enjoy." These are simply transla-

THE KIND OF PERSON I AM

Today's Date

		Priority
Immediately		
One Week		
One Month		
Three Months		
Six Months		
One Year		
Two Years		
Five Years		
Ten Years		

THE ACTIVITIES I DO

Today's Date

		Priority
Immediately		
One Week		
One Month		
Three Months		
Six Months		
One Year		
Two Years		
Five Years		
Ten Years		

THE WEALTH I ENJOY

Today's Date

		Priority
Immediately		
One Week		
One Month		
Three Months		
Six Months		
One Year		
Two Years		
Five Years		
Ten Years		

tions of the Be-Do-Have desires we've listed previously. Down the left hand side of the worksheet you'll notice different time horizons. Start by putting a date the top of the page so you're clear on the time horizons. You can even specify what the exact date will be in 1 month, 6 months, etc. down the left.

DEVELOP YOUR GOALS AS AFFIRMATIONS AND PRIORITIZE THEM

In transforming your desire list to a formal goal list you'll want to use proper syntax. Specifically, you'll write your goals in the first person present tense, describing the goal as if it were already achieved. Simply tell the truth in advance. For example the desire "I want to have a 1000 square foot nine station salon in Roxboro Mall in 2 years" ultimately gets translated to "I own a 1000 square foot salon with nine stations in Roxboro Mall" and placed alongside the 2-year horizon.

Now, look at all the A priorities on your desire list. What we're going to do is rewrite them and place them on the sheet of paper under the appropriate heading and across from the appropriate time horizon. Naturally, even among the A priority goals you can further differentiate with A1, A2, A3 and so on and it's a good idea to keep that clear. After you've transcribed your A list move on to the B list and C list and what you'll have is a complete written goal list of all that you set out to achieve.

Keep in mind that goals and priorities change. It's a good idea to review your entire goal list regularly to make sure it's in harmony with your then-current thinking.

HOW TO APPROACH GOALS IN A PRACTICAL SENSE

TRANSLATE GOALS INTO A PLAN OF ACTION

Major goals need to be broken down into smaller goals. Each of those smaller goals will give birth to a list of specific actions. You can't do goals, but you can do activities. Starting with your most important short-term goals first, you'll want to develop sub-goals and then for each subgoal you'll formulate an activity list. It's just a practical approach to start defining what you want to accomplish. Next, you devise a specific plan of action. Finally, you'll formulate exactly what you'll need to do to bring your plan into action.

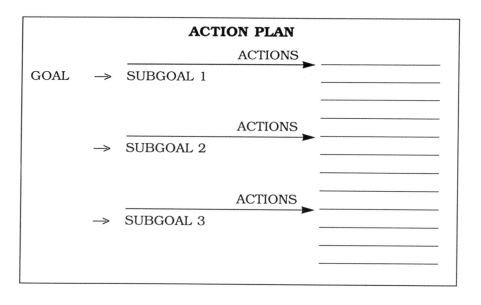

Let's say you have a goal to earn $50,000 a year. It's certainly achievable! The first way to approach it is to break it down. Let's say for ease of discussion, that you work 50 weeks a year and are paid a 50% commission. You'll need to get your weekly service revenues up to around $2000. If you're at the salon 5 days a week, that translates to $400 in services a day. If you work 8 hours per day, that means $50 an hour in services. That means that you need to be doing a lot of chemical work and maintain about 100% productivity. Can you do it? Yes you can!

The next step is to calculate where you are right now. Review your performance over the last month. How many dollars worth of service did you provide? Divide that by the number of hours you worked. What was your service income per hour? How does that relate to the $50 an hour goal?

Next, devise a plan of action to move your income from where it is to where you want it to be. A $50,000 a year income goal is probably going to involve activity that creates and retains new clients, that stimulates additional client purchasing and that improves time management and efficiency. In reading this book you're getting ideas on all of these topics. In fact, reading this book is one specific activity you're doing to achieve a financial goal. I've found that education always pays for itself and is always worth the investment of time and resources. As someone once said . . . "if you think the cost of education is high, try the cost of ignorance!"

Continuing with the $50,000 income goal, decide on some specific and measurable things you're going to do consistently to attract and retain new salon guests, to enhance guest spending, and to maximize your time efficiency. Let's say that one of many measurable strategies that you're going to implement to increase client spending is to start charging for conditioning treatments. Keep in mind that this is just one of many different things that you'd be doing to achieve your goal, but it's a good example because it illustrates many of the principles in action.

Your activities surrounding that particular subgoal could include:

1. Calculate what percentage of your guests are purchasing conditioning treatments right now.

2. Work alone, or with your manager and associates, to develop a menu of perhaps six or eight treatments. (This activity itself could have several steps to get it accom-

plished like scheduling a brainstorming session, preparing menu copy, having a review session, designing the graphics, then having a printed version of the menu created.)

3. Develop and practice a short presentation for each treatment.

4. Set specific and measurable goals for increasing the percentage of guests who are purchasing conditioning treatments. (Lets' say, for example, you calculated that only 3% of guests were buying treatments. You might set a goal to have 30% of them purchasing treatments within 90 days—I think you could probably do it in 30 days with focused effort, but let's use 90 days for the example. In month one you want to be at 10%. In month two, 20%, and by the end of month three you want to be at 30%.)

5. Get into action and start stimulating the impulse purchase of treatments. You will receive many specific techniques for increasing purchasing in Section III.

6. Evaluate your performance and refocus your energies. As days and weeks go by you'll want to calculate how you're doing. If your numbers reflect better than expected performance then revise your treatment goals upward or speed up the time frame. On the other hand, if progress is sluggish, see what you can do to get up to speed. (For example, it's not at all unusual to revise and refine your service presentations to maximize the percentage who say "yes" to your impulse purchase ideas. This can only be effectively done when you're in the thick of it. So you must fearlessly proceed and give yourself permission to make a few blunders on your path to success and then fine tune your approach as you go along.)

FOCUS ON RESULTS

Remember the 80/20 rule, where in any list of activities you'll get 80% of your benefit from 20% of the activities. Of the six activities just listed for enhancing treatment purchases, which is the payoff activity? It's number 5. So, here's the key—don't let yourself get bogged down in the first four. As a matter of fact, you'd probably be best off starting with number 5 before the first four are even done! You can start on number 5 with your very next visitor. You don't have to wait for 2 or 3 weeks until the menu is developed. You can start enjoying the fruits of this goal at once.

This point is worth emphasizing because for you to make rapid and measurable progress toward your goals you want to put your major effort toward those things that are going to make the difference. Many people let themselves get bogged down in the minor details that make little difference to the bottom line. Avoid that trap. Make it a point, whenever you develop a list of activities for any goal, to zero in on the payoff steps and make a bee line toward working on them.

Now that you've got a good feel for how to break goals down into subgoals and specific activities for their accomplishment you can start to make a big picture plan for your future. You know what you want to do, who you want to be, and what you want to have. Next break those desires down into subgoals and then bite-sized tasks.

You may find it productive to invest an entire afternoon in designing your future. Get out all your goals and begin brainstorming all the things you'll need to do for each. It's a real adventure. It's also self-affirming and motivational. You'll develop a list of subgoals and tasks for each of your major goals. Have a separate sheet of paper or a separate folder or binder for each. Develop to-do lists for each goal. You'll find that activity lists will become a major tool in your goal management. Prioritize the activities. Give each activity a deadline. Make your goals a daily reality.

WAYS TO KEEP YOUR GOAL CONSCIOUSNESS FRESH

Scheduling goal-achieving activities each day is something we're going to go into some detail about in the pages ahead. That's the practical reality of success. Additionally, we must keep ourselves motivated. Our goals are a fantastic motivational tool and here are some ways you can use your goals to stay on fire each day.

1. *Review your goal list often.* Place your major goals on an index card, remembering to word them in the first person, present tense as if they've already been achieved. You may want to make two or three copies of that card so you can have it with you all the time. Tape one on your bathroom mirror. Have one in your car. Keep one in your wallet. Then make it a habit to review your goals each and every day and several times a day. This daily process helps tremendously in keeping you focused.

2. *Proclaim your goals regularly.* Another good idea is to say your goals out loud as affirmations. Say them as if they've already happened. Say them first thing in the morning, at lunchtime, and before bed at night. Say them with all the emotion and passion you possibly can. Put a lot of expression in your voice.

3. *Visualize your goals as having been achieved.* Picturing your goals in the mind's eye is a great technique. Imagine yourself with your goals already being achieved. I've found it beneficial to act as the producer, director, and star of my own little mental film that I play for myself each morning before I even get out of bed. You can use this as a way to visualize what you want as if it's already been accomplished. Make your little film rich in detail. What are the surroundings? Who are you interacting with? How do you feel? What are people saying? How is everybody acting?

4. *Experience your goal realizations in advance.* Act out your goals in advance. Literally experience their achievement in advance. When I was a teenager, I sometimes went into my

room alone and played my favorite music and imagined that I was the singer. I saw the audience. I sang with great emotion and feeling. I accepted the applause of the audience. I was there! I'll bet you've done the same sort of thing yourself as a musician or athlete or celebrity. You can use the same strategy to make the accomplishment of your goals more vivid. Put yourself there in advance. Play act a little. Make yourself the platform artist or the hairstylist appearing on your favorite talk show. Play the role of the artistic director or the stylist at a high-fashion photo shoot. Put yourself wherever you dream you want to be and then experience it in advance. It's a potent way to make your goals more present and can help to convince your mind of the reality you want unfolding in your life. Make each day count.

By the way, these methods not only work to "massage" the big picture goals, they can also be used effectively for subgoals and daily activities. It's not just about where you want to be in 5 years—it's about what you need to do accomplish today to get where you want to be in 5 years. You can affirm and visualize and review and experience what you want to accomplish today! Review, proclaim, visualize, and experience what is on the agenda for today. Don't for a minute forget the overreaching power of prayer and meditation. If you believe in a higher power don't neglect to reaffirm your personal commitment and be sure to ask for guidance and assistance in performing your daily mission.

You can start off the morning by visualizing yourself recommending conditioning treatments with confidence and persuasive appeal. You can experience the client's positive response to your prescriptions and delight at the results. Furthermore, you can write down on an index card "I will sell four treatments today" and say that aloud to yourself on your way to work, when you're in the washroom and during your breaks. That will sure help keep you focused!

Give today your all. There's an old saying that yesterday is a cancelled check so don't count on it; and tomorrow is a promissory note so don't count on it; but today is ready cash, so use it and spend it wisely.

You create your future each and every day. Your future, whether or not your goals and desires will come to pass, is completely dependent on your thoughts, attitudes, and activities each and every day. Your future will be the sum total of what you do today, tomorrow, the next day, next week, next month, etc. But it all starts with today. The only thing you have any direct control over is what you do today. Actually, the only thing you can really control is how you handle this very moment.

Hold yourself to a high standard. Have the discipline necessary to meet the requirements of the moment and the challenges of life as they unfold. Affirm to give it your best at the start of each and every day. Then do the very best you can. At the end of the day make note of what you accomplished. How many steps were you able to take toward your goals? How was your purpose served? How can you do better the next time? Remember that tomorrow is a fresh day and you get to make a fresh start.

TIME MANAGEMENT ESSENTIALS

There's one factor that's the great equalizer. That factor is time. Whether you're rich or poor, young or old, short or tall—when it comes to time there are no favorites. We all have exactly the same amount of time each day—24 hours, that's it. No one has figured out a way to make the day any longer! No matter who you are you get the same 7-day week and the same 52-week year.

You can never get more time and you never really know how much you have to begin with. Time is your most precious resource and it's being constantly depleted. One of the major factors that separates the highly successful beauty professionals from their colleagues is how they manage their time.

Let's review the eight golden laws of time management for salon professionals. By mastering these fundamentals you'll be able to experience much more productivity and move along much more rapidly toward the realization of your goals. The difference between someone who works 8 hours a day for 5 days a week at the salon and earns $20,000 a year and one who puts in exactly the same number of hours in exchange for $50,000 is all about what they do with their time. Let's proceed then to discover the secrets of success.

LAW I: SCHEDULE YOUR TIME

Many of our income-building goals are going to involve perfecting and repeating the same process over and over again with each salon guest. Once we have our money-making "cookie cutter" in place, we role out the dough and go for it. Naturally, we always try to improve speed and efficiency. Although there are practically always a few subprojects we must schedule to get our various "cookie-cutters" going and operating smoothly, it's mostly about maximizing performance and measuring results over time.

However, as we've discovered, many major goals will yield numerous subgoals and lists of activities. For example, the $50,000 income goal will probably involve the development of several "cookie cutters" and it's appropriate to have a clear picture of all the ingredients you'll be bringing together before you start. That way you have a planned approach that will be much more effective and efficient.

The first principle of management is planning. As Dale Carnegie said "plan your work, then work your plan." Know what you're going to do and when you're going to do it. Then, commit those activities to a specific schedule and put them in your calendar of activities with a specific day and time that they will be accomplished.

It's a good practice to start developing "to do" lists each night for the next day. A weekly "to do" list is also a great idea and I do mine on Sunday night so that I can start the week off and running! Your lists will contain activities large and small so prioritize them and include a rough estimate of the time you'll need to get each done.

Next, schedule your activities in your daily activity planner. In addition to your booking sheet in the salon, it's a good idea to maintain a separate personal schedule book for these and other activities. You can get one at any stationery store. Some of these "time management systems" can get complex and I've always found a simple weekly planner worked just fine.

For efficiency, group similar activities together. Make all your phone calls at once. Run all your shopping errands at once. Do all the paperwork at once. Cross items off the list as they're accomplished. The process of crossing items off the list will give you momentum.

At the end of the day, make tomorrow's list. At the end of the week, make next week's list. You may have to transfer some undone activities from one list to the next, but try to get as many done by the deadline as you reasonably can.

LAW II: MAKE TIME

It's not unusual to have an experienced designer or salon owner/manager so fully booked as to feel a time crunch. They would like to be able to invest time and effort into long-term plans but are so occupied serving clients they barely have enough time to breathe. What do you do then?

If that's your situation, let me first of all congratulate you! A price increase is probably in order, which should reduce demand slightly and free up a bit of time at least. Beyond that, you must have the discipline to schedule time away from clients to major in the majors. Have the self-discipline not to relent and book service appointments during times you had planned to be occupied with other projects. Sometimes the lure of a few extra dollars puts priorities on the back burner and then scheduled activities don't get done. It's a mistake. The thing you have to remember is that in our world of competing demands sometimes you have to drop a short-term nickel to pick up a long-term dollar. It takes discipline and commitment to keep the urgent demands of the moment from lousing up your scheduled "big picture" plans.

When your time is particularly sensitive, don't make the mistake of skipping the planning step. Furthermore, you'll want to find ample slices of time when you can make quality progress. A half-hour here and there is fine for small stuff, but a regularly scheduled full or half day devoted to major projects

is often in order. That way you'll make sure the time you
yields the most results.

LAW III: RESPECT YOUR TIME

There's never any free time for the goal-oriented person in the
salon. Taking time to read dime novels or complain in the staff
room is not how people going places spend their most precious
resource.

Your time is more valuable than money. You can get more
money but you can't get more time. And once time is spent
there's no turning back the clock. You wouldn't let someone
steal your money before your very eyes. Don't let other people
divert you from your schedule of activities with time-wasting
chatter and nonsense.

"Down time" is the time to work our "big picture" plans. It's
also the time to perform the support activities that further our
goals. For example, if you do have a fully booked schedule, you
can always cultivate new salon customers and retain those that
you have. Down time is when we put together our promotional
and networking campaigns and get them started. Down time is
when we make follow-up telephone calls and write thank you
notes and reminder cards. Down time is when we prepare displays and presentations. So, down time is not down time. It's free
time to invest in our future performance.

Don't let others waste your time. Whether it be coworkers
who ambush you to ramble on about themselves or clients who
want to linger endlessly, you must have a strategy to stay in control of your time. It's best to avoid these traps to start with. Interestingly, folks usually have more respect for the time of people
they perceive as busy. So simply staying busy is a great place
to start. However, sometimes you have to politely excuse yourself. Bow out of long conversations by saying "It would be fun to
chat but I've got myself scheduled with something that I can't put
off any longer so excuse me."

Also, discourage interruptions when you're up to your
elbows in a project. Interruptions can only slow you down. That

means not taking telephone calls. That can mean being away from the salon and in a quiet space to get projects done. I can remember the days when I kept an office at one of my salons. There was just so much going on that interruptions were constant. Eventually I created an office in my home so I could get some work done. I know many salon owners and managers who have done the same thing to great benefit.

LAW IV: INVEST YOUR TIME

Occasionally, a cosmetologist might say "I don't get paid for doing those things." Let me hasten to say that such an attitude does little to cultivate success. A primary principle of achievement is to do more than you get paid to do as an investment in your future. It's also known as the investment principle that any financial planner will tell you about. At first, you invest $10 and get only a $1 return. But if you do that regularly and constantly the accumulation of those $10 deposits yields a return greater than $10. So, even though you're still just putting in $10 during the time period, your accumulated return is greater than $10. As considerable time passes even though you're putting in the same $10, you're experiencing an accumulated return of $200, $400, $600, or even $1000. That's the investment principle.

This principle of return has been known since Biblical times. It doesn't hold true just for money. It also holds true for intelligent effort. You can replace that $10 in the example with the idea of "sweat value" also known as extra effort. So that extra hour of effort that only yields a small return today can surely grow. Put in that extra effort regularly and consistently and you'll be surprised how the investment principle will come into play over time. At first, you work harder that what you get paid for. As time goes on you get paid for far more than what you do.

This is not about anybody exploiting anybody or taking advantage of anyone. To think that way is negative and counterproductive. Philosophically, you want to "sow" your extra effort with happiness and joy. You deserve better than negative energy tied into your invested effort. You want sweet results so plant sweetly. A lot of bitterness, anger and resentment will affect your eventual harvest.

LAW V: ENERGIZE YOUR TIME

Develop a sense of urgency. Be relaxed, but move along rapidly and quickly. Be productive. Get on with things. Constantly ask yourself the question "what is the best use of my time right now?" Don't fritter away your time on minor things. Focus your efforts on what will yield the most productivity and profit for the time spent.

Condition yourself to have endurance and quickness. Being in good physical condition will give you more energy and will enable you to work longer and faster. An exercise program is not only good for your health but for your overall productivity. Furthermore, watch your diet. Some foods have a tendency to weigh us down. Be especially careful of heavy lunches—they can completely slow you down for the rest of the day. Try fruit or a salad instead of cola and fries. You'll find that you feel better and can work faster as a result.

Work with your natural cycles. Some people are morning people. Some people are night people. We all have different energy levels at different times. Be sensitive to this and try your best to schedule your activities during times that take advantage of your energy cycles.

Be mentally prepared to work effectively. Get yourself in the right frame of mind so you can focus on productivity. This is where your goal lists and affirmations can come in handy. Have discipline. I find that a morning routine including a bit of exercise, inspirational reading, and meditation is an ideal way to prepare the mind and spirit to make the most use of the day we've been given.

Keep yourself on track throughout the day. Don't let one unpleasant episode or stroke of misfortune ruin an entire day. It's easy to lose positive energy and focus when confronted with less than ideal news like a sudden cancellation or news that a service needs to be redone. How about starting off the day with the morning mail that includes a bounced check from a customer being returned by the bank? They're all downers. But don't let them sap your positiveness and kill your day. Respond to these disruptions promptly and get on with the positive news of living. You must have the self-discipline to keep yourself energized all day. Remember the old ditty:

> Whether the weather be fine, whether the weather be not. Whether the weather be cold, whether the weather be hot. We'll weather the weather, whatever the weather, whether we like it or not!

LAW VI: ENJOY TIME

The salon is a fun place to be. Enjoy yourself. Have a good time. I can't think of a more enjoyable place to be than a salon. It's a relaxed and creative environment loaded with interesting people who want to have a good time.

Remember, work during work time and relax during leisure time. Be where you're at when you're there. Avoid losing focus and thinking about play when you're at work. Avoid losing focus and thinking about work when you're at play.

So often we hear about "workaholics" who just keep themselves busy all the time. That's when work is abused and used as an escape from other issues of life. Some of us have been taught to think of hard work as virtuous—and it is. But so is family life. So is good clean fun.

Owners and managers in particular are subject to enslaving themselves to long and often unproductive hours of overwork. The joy goes out of business. They lose a sense of happiness and it becomes drudgery. When we overwork our productivity nosedives, our creativity goes into freefall, our enjoyment vanishes, and we risk burnout! Then we're no good to anybody. Pace yourself for the long distance journey.

Part of enjoying your time is learning to say no. Avoid making commitments that will overload your plate. The desire to do it all, be it all, and have it all may appear great, but the reality is often that nothing is done very well or enjoyed very much. Focus on quality not quantity and don't feel you automatically have to say yes to some project or activity just because it's offered.

LAW VII: USE TIME EFFICIENTLY

Time management experts universally recommend the idea of "single handling it" and "avoiding clutter" as basic tenants of efficiency. So often we have a tendency to put tasks off into secondary time zones rather than handling them the first time they come up. Only handle an item once and when the task is in process finish it completely. That means when the perm is done we don't leave the rods in the sink all afternoon. We tend to it immediately.

Contemplate salon cleanliness, for example. It's something we all have to play a part in. We have two options. Clean as you go or let it accumulate and clean it all at once later. Why let a small thing grow into a major project when all it requires is paying attention and making a few simple extra movements as we go through the day. Plus, when you think of the price we pay in terms of professionalism it just plain isn't worth it!

Letting clutter accumulate at the front desk, at the back bar, or at the station has more slowing effect than anything else I can think of. When we allow it to happen we're constantly rummaging through stuff to find what we want and moving piles of clutter from one spot to another without doing anything about it. It's maddening!

The solution is to finish tasks completely before moving on to the next step. That way we avoid leaving clutter in our wake. One of the great wasters of time is taking a simple task and picking it up and putting it down and picking it up and putting it down and ending up with a half-dozen things half done. Neither the consumer nor the coworker wants to muddle through someone else's mess. It not only slows things down, it also makes one's thinking take pause!

F.W. Taylor, one of the founding fathers of Industrial/Organizations psychology devised the concept of time/motion efficiency a century ago. It's another idea for using our time more efficiently. Taylor's idea was to make as few movements and use as little time as necessary to complete a task. It makes sense when you think that a few seconds here and a few seconds there add up to minutes and hours. It was based on his theories that the assembly line began to dominate manufacturing. Now, though we don't work assembly line style, the fact that your time is money makes it practical to figure shortcuts and ways to do the same thing with less movement and time.

Have you ever seen a waitress at a restaurant race around making several trips to the kitchen when with a little forethought everything could have been done at once? I feel sorry for the poor soul—this running around unnecessarily. The same thing happens in the salon. Take something as simple as visiting the dispensary for supplies. I've seen designers rush from pillar to post to pick up one item only to race back to get another item from exactly the same spot. We don't need to do that to ourselves! A little proper prior preparation can significantly reduce the time and motion we're using and make us more efficient. The idea is to move along quickly without appearing to be rushing.

LAW VIII: AVOID PROCRASTINATION AND DEVELOP PERSISTENCE

You know the old saying that procrastination is the thief of time. Procrastination is about putting things off to a secondary time frame when they could just as easily be acted on right now. This can happen with basic things like writing thank you cards to important things like starting a major project. Procrastination burns up more hours and days that you can possibly imagine. So often when we procrastinate we end up just sitting idle.

"I do it now. I do it now." That's the mantra of success. "I do it now." Not doing it now is actually painful when you stop and think about it. There's guilt. There's the nagging feeling of wasting time. There's the feeling of not being in control. The fact is that putting things off is actually psychologically stressful and often has physical symptoms. Why put yourself through it?

The opposite of procrastination is action. You can have all the goals and affirmations and good intentions in the world, but if you don't follow through with concrete action you have nothing. Plant your feet firmly on the ground where you are right now and commence your journey. Move. Do something. Do anything. Start. Don't just think about starting—get started! The present moment is the only thing you have any control over so use it with value.

Sometimes well-intentioned people have a hard time shifting into action. Why? Sometimes people don't shift into high gear because they're lazy. There's no sense in sugarcoating this one—let's call it what it is. Sloth is one of the seven deadly sins. It's not so much that they can't do or are afraid to do it—it's simply that they won't do it! Other than dealing with improving physical energy, which we discussed earlier, the main reason for laziness is a plain old lack of discipline and mental toughness. People are complacent when the prospect of changing appears to require more effort than it's worth. If you've read this far into the book, I think you're somewhat motivated to improve your life and career and that desire in itself will eventually defeat laziness because the pain of staying where you are is greater than the effort associated with growing. However, don't make reading this book merely an intellectual exercise—act and start right now right where you are.

The scope of a project can sometimes appear daunting. Often the major goals people want to accomplish in life they put off because they don't know where to begin. This kind of procrastination, though common, is easily addressed. It's really quite simple when you realize that all you have to do is break the big project down into several smaller projects and then break those down into bite-sized chunks that can be put on a schedule. The main thing is to get started and let momentum build. Then, you'll be surprised to discover how quickly the whole things takes form and how easily your activities can be streamlined. As Henry Ford said "nothing is particularly hard if you divide it into small jobs." Promise yourself "I'll work on this for 10 minutes" and you could end up with several hours of meaningful effort merely because you were able to get started.

Sometimes people don't get going because of fear. Fear of failure holds people back. That's one of the reasons why so many give up on their dream at the first sign of difficulty. They start with a half-hearted fear-ridden attempt and then take a little stumble and just throw in the towel. That, however, is not the way successful people move forward.

Here's a little dose of reality—anything worthwhile that you will try to accomplish in life is going to have challenges and difficulties. The more significant the goal, the more obstacles will be in your way. Successful people are realistic about this in advance. They are determined and promise themselves to do what it takes to get to their goal.

One of the laws of success is that you must put forth all the effort, energy, and risk in full and in advance before the success will be yours. You've got to make it through the challenges and difficulties. Remember the old ditty

> When things go wrong as they sometimes will.
> When the road you're trudging seems all uphill.
> When funds are low and debts are high, and you
>   want to smile but you have to sigh.
> When care is pressing you down a bit—Rest if you
>   must, but don't you quit!

This issue of sticking with it is so important because the other option, giving up, is failure. It's accepting failure. It's sad that many people develop a habit pattern of giving up before they've really even started the hard work. It's a habit pattern that will enslave anyone to a life of mediocrity and "if only's." That is one of the reasons people are afraid to start to begin with, because their pattern is to give up early in the game. So, why bother getting started if they're going to give up before success anyway?

My favorite quotation—a classic—on the topic of persistence was spoken by President Calvin Coolidge when he said:

Nothing in the world can take the place of persistence. Talent will not; nothing is more common than the unsuccessful man of talent. Genius will not; unrewarded genius is almost a proverb. Education will not; the world is full of educated derelicts. Persistence and determination alone are omnipotent. The slogan "press on" has solved and always will solve the problems of the human race.

Experience teaches that the largest measure of success is obtained simply by hanging in there while others drop out of the race. You'll find in life that a lot of people will simply give up. Surprisingly, it can be the most articulate, intelligent, and attractive colleagues who quit prematurely. Having staying power counts more than any other factor. That's why desire is so crucial. No one is going to "hang in there" unless they really want it. Furthermore, that's one of the reasons why comprehending your purpose is so vital. In times of darkness and challenge, your sense of purpose may be the only thing that sustains you. Avoid procrastination and cultivate persistence.

SUMMARY

The first principle of self-management is planning. There's an old saying that if you fail to plan, then you plan to fail! Transforming your desires into goals and your time into results starts out with planning and ends up with performance. Here's what we learned:

- Translate your desires into specific written goals with time horizons.

- Figure out an objective measurement you can use to gauge your progress and develop action plans.

- Keep yourself mentally focused and psychologically motivated in the direction of your goals.

- Have discipline over your time and follow the basic laws of making and keeping your schedule and performing with efficient work habits.

- Develop a sense of urgency, avoid procrastination, and persist until you get where you're going.

The race is not always to the swift, but to the one who keeps on running. You can reach your goals. As the old saying goes "inch by inch it's a cinch." And worry not, there's plenty of room for you at the top. The more success and prosperity you enjoy as an individual the better it is for the industry. You'll discover that those who journeyed the road of success ahead of you will be more than happy to help show you the way. After all, they want more company at the top—it can get lonely up there!

PART II

MASTERING THE PSYCHOLOGICAL COMPONENTS:

Influence Client Psychology and Stimulate Your Own Psychological States

C h a p t e r 4

Essentials of Client Psychology

66

You can have all your heart's desires if you help enough other people get their hearts' desire.

99

Why do people go to a salon for services? Is it because they need the haircut? It can't be that because every day we see people walking down the street who need and can afford a haircut or a facial or a manicure, but yet don't obtain it. Is it that they don't want to spend the money? Is it that they don't want to take the time? The haircut, or any other salon service for that matter, is not an issue of fundamental need. The fact is that there are other motivations, psychological motivations, that dictate when and how a person will engage salon services.

WHAT YOU WILL DISCOVER IN THIS CHAPTER

- You'll understand how dissatisfaction is at the root of salon service demand.

- You'll come to grips with tremendous impact appearance has on quality of life.

- You'll learn about the psychology of grooming and how to provide clients with an ongoing and uplifting afterglow to their visit.

- You'll discover how to use specific communication strategies that will enable you to make a more dynamic emotional impact on salon guests while making your consultations more influential at the same time.

IT'S WHAT THEY WANT THAT COUNTS

Clearly, the population wants salon services. They've proven that with their pocketbooks since the days of Cleopatra. We know that it's not because they need them. They don't. Ultimately, the important question to reflect on is "why" they want salon services. The answer to that question appears to be twofold.

1. They want to make a good appearance. They want other people to accept them. At the very least they want to be suitable, presentable, and acceptable to others. This relates to their basic want for social acceptance and fellowship. Of course, for a large percentage of people the want is more than to be merely acceptable.

2. They want to feel a certain way about their appearance. They want to be recognized for their good appearance and even experience feelings of prestige and leadership. Their sense of identity and their very personality is inti-

mately linked to their appearance. How they look is a tremendous source of self-expression.

THEY'RE BUYING THE IMAGE THEY WANT TO PROJECT

This psychological aspect of appearance is the linchpin of the entire beauty business. Ultimately, we are in the business of delivering to people the appearance image they want. The means by which we accomplish this are cutting, perming, relaxing, and coloring hair, and the whole array of nail, aesthetic, and cosmetic services that we offer.

What we are providing is image. That's what consumers are buying. They're buying the image they want. They're not buying the haircut; they're buying the image that they hope the haircut will deliver. All high-achieving salon professionals I know understand and believe this reality. Consumers want to make a good appearance and they want to feel good about the appearance they make, and ultimately that's what brings them in the salon door.

It's their image and creating it is an art. Always keep in mind that it's what the client wants that's most important. We perform our art on living, breathing human beings. They have to feel absolutely comfortable with the image we're creating for them. It's crucial that we be sensitive and vigilant about helping clients radiate the image and appearance that reflects their personalities and tastes. The art is our ability to bring together what clients want— the raw material we have to work with and a sense of fashion and style. And it is an art! (Fig. 4-1)

We know that this is true because you can take exactly the same hairstyle and put it on two different people and have it all right for one and all wrong for the other. If clients aren't comfortable with their new style, they'll speak up about it loudly, make no mistake. If they don't like the color, cut, or curl of their hair, they don't feel like themselves. Our goal is to help them express themselves fully. That's what they want and it's what we should want and work toward for them.

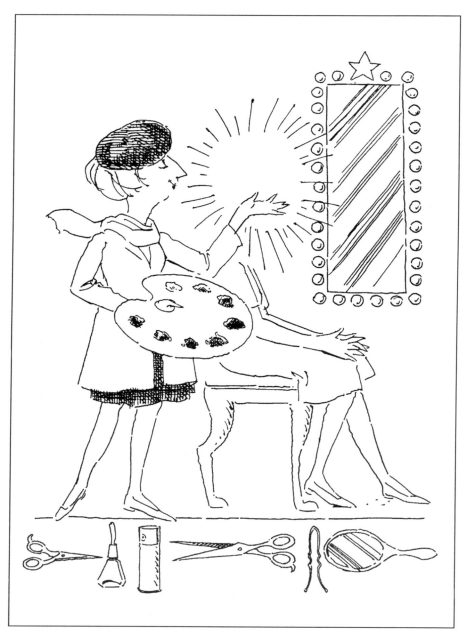

Figure 4-1 The art of cosmetology is the ability to combine the client's desires, the raw materials at hand, and a sense of fashion and style to create a pleasing or provocative result.

THE PSYCHOLOGY OF COSMETOLOGY

SALON GUESTS ARE DISSATISFIED

By the time consumers get to your salon, they're dissatisfied with their image. They don't have what they want. They're visiting in hopes of receiving the image that enables them to feel more fully actualized. When their dissatisfaction is more painful than the expenditure of time and funds, they come to the salon.

But here's another point to keep in mind—in general people are dissatisfied with their appearance. Take hair for example. If you took a poll you'd quickly discover that a large majority of people are generally dissatisfied with their hair. If their hair is straight, they'd like it curly. If their hair is curly, they'd like it strait. If it's brown, they'd prefer blonde. If blonde, they'd like it red. Shorter, longer, thicker, thinner—it goes on and on! People are dissatisfied with what they have and want something different. Often, what they have, and what they've settled for, isn't what they really want!

This is good news for the cosmetologist, especially on a client's first visit and at other crucial times a for client as well. We have the opportunity to deliver the image that the client really wants. For those first-time guests, most of all it's important to keep this fact in mind—they're with you because they hope you'll be able to deliver what they want. They hope you'll achieve it more closely than the last cosmetologist they visited. As a matter of fact, if you've been recommended by a friend their hopes can be particularly high. They're ready for a change!

ALWAYS COMMUNICATE DELICATELY

It's important to avoid criticizing the work of the previous cosmetologist. It's just not good professional form. It is also potentially highly insulting to your guest. Keep in mind that it's not only the other person's work you're talking about—it's the client's hair you're talking about!

I've seen designers comb a new client's hair around and then flop it over her face, going on and on about what a lousy job was done. This makes the client feel about one inch tall and is just extremely counterproductive communication. If it's absolutely necessary to say something or if the client is making negative remarks about previous cosmetologists then be very low key and take the higher road. You could say "I'm sure they did the best they knew how. I have a different opinion on what will work best for you and I think you're going to like it very much." Enough said!

APPEARANCE DISSATISFACTION IS PSYCHOLOGICALLY CRIPPLING

How people feel about how they look has a tremendous impact on their experiences in life. You know that this is true from your own experience. Much of the psychological impact appearance has on us can be gleaned by examining our feelings about our appearance when we interact with others or look in the mirror.

These feelings can be deciphered by listening objectively to our "self-talk." We carry on a nonstop internal conversation with ourselves; this internal dialogue is our "self-talk." It reflects our state of mind and our feelings and attitudes about what we're encountering at the moment. Much of this "self-talk" has to do with how we feel about how we look.

You'll be surprised to discover that from time to time you probably say some truly awful and negative things to yourself about how you look. So bad in fact that another civilized human being would never dream of leveling the put-downs and insults that you give to yourself from time to time. You're not alone—this is a most common phenomenon.

Let me prove it to you using an example I often relate during my Supernatural Salon Income Seminar. Imagine waking up in the morning after a night of hearty partying to discover yourself disheveled and a bit bleary eyed and incoherent. You need a cup of coffee so you stagger to the kitchen only to open the cupboard and discover you're fresh out. Being a day off from work, and planning to lounge around the house all day, you decide to make a quick dash to the local supermarket to run in and get a package of coffee to bring home. You're only going to be gone a few minutes so you throw on a jacket and decide not to even bother with your tangled mess of hair.

Let me ask you, have you ever gone out in public like that before? We all have. So you walk into the supermarket, trying to keep a very low profile, and what always seems to happen? You run into somebody you know! Heaven forbid that it be someone you'd want to impress socially—or that old boyfriend or girl-friend you'd only want to have see you in the best light possible to make him or her feel jealous or regretful at the break-up of the relationship. But instead you're a mess! You may try to play aisle tag in an effort to avoid the person. You may even scamper out of the store like a wounded and frightened animal. But, chances are that you'll be unable to avoid making contact (Fig. 4-2).

What's your self-talk then? What are you saying to yourself as you engage in conversation under these awkward condi-tions? We've all been there. I'll tell you for sure that you're tear-ing yourself apart. You're completely preoccupied with your appearance and unhappy at how you're presenting yourself. You're wondering how the other person must be judging you. You figure the individual is thinking all sorts of critical thoughts. You speculate that this could affect how the person deals with you in the future. You're concerned about how it will influence the person's opinion of you. You contemplate what the individual might say to others about this. You feel almost as if you need to make an excuse for yourself. It's one big negative! Because you're so preoccupied with all the negatives, you're certainly not in a state of mind to be open or effective in your communication. That just makes matters worse!

APPEARANCE DISSATISFACTION COMPROMISES QUALITY OF LIFE

Brain researchers have discovered that our conscious mind is only capable of thinking one thought at a time. Well, if we're expe-riencing negative thoughts about our appearance are we in a position to be experiencing happiness and joy? Of course not.

Figure 4-2   When your hair doesn't look good, your ability to interact with others is compromised and your self-esteem takes a nosedive.

Our appearance has a great deal to do with how we feel about ourselves, especially as it relates to our needs for social acceptance and community. If we're not happy with our appearance, we fear rejection and condemnation and ridicule. You bet this influences our quality of life! It has an impact on our whole experience of life! If we're so uncomfortable with our feelings about our appearance we may begin to turn anger and ridicule inward toward ourselves—and literally begin to hate ourselves. When we reject and disown ourselves, we're in the midst of a tremendous amount of unhappiness.

Self-image has to do with how people look at and evaluate themselves. Of course, many elements make up how we see and feel about ourselves at any given moment. Yet how we feel about our appearance (skin, hair, clothing, weight, etc.) is absolutely fundamental and has the greatest impact on how we feel about ourselves as people.

There's been a lot of talk about "good hair days" and "bad hair days" and it's true. Of all the elements making up how people feel about themselves at any given moment, how their hair looks has a pivotal impact on how they experience the day. Hair has a greater impact on how people feel about themselves than does their body shape and size, the amount of money in their pocket, the clothes they're wearing, the career they're pursuing, the car they're driving—more than any of them. Amazing but true!

And this is not only true of women. This is also very true of men. If you've ever had an opportunity to see one of those late night infomercials on men's hair replacement, you can relate to what I'm communicating. When a man experiences some hair loss, this is a traumatic situation. It can have a devastating psychological impact. In fact, a man can lose his entire sense of identity and his entire self-image can undergo a personality change. The whole experience of life is affected. His sense of confidence and ambition, feelings of worth and attractiveness, and sense of happiness and joy all can vanish with hair loss. It doesn't even have to be dramatic hair loss. The impact on his life can be devastating as he begins to convince himself that his social, romantic, and career ambitions have slipped away.

Thousands of men have been motivated enough to spend thousands and thousands of dollars to cosmetically cover up the hair loss with "hair replacement systems." Thousands of others have spent tens of thousands of dollars on surgical hair replacement. This kind of discretionary income isn't being spent frivolously. Men are seriously concerned about the state of their hair.

Medical patients, too, feel the impact of their hair. Sometimes people have sickness or undergo treatments that cause hair loss. This becomes a major psychological issue that can cause severe and chronic depression. That's one of the reasons that insurance coverage will often include custom-made wigs for people devastated by medically related hair loss.

To a significant degree, a sense of personal identity is intimately connected to a person's hair and his or her feelings about it. On it's own strength, how it appears can literally make the difference between happiness and depression—between confidence and fear—between self-affirmation and self-loathing.

APPEARANCE SATISFACTION IS PSYCHOLOGICALLY UPLIFTING

Consider the other side of the equation. Look at the impact that "good hair" has on people. Think about a really "good hair day" you've had recently. Perhaps you were going out for a special evening. Perhaps you had an important business appointment or were going to a social gathering, a wedding, or a reunion. That day the blow dryer was working magic for you. Every hair fell into place perfectly. As you finished your style with some gel or spray, you achieved exactly the effect you were looking for with every hair perfectly in place. It's almost as if the bone structure of your face came to life. Your skin took on a new glow. It's as if suddenly you looked 10 years younger or 10 pounds lighter. You know the feeling!

When you walk out the front door and leave the house on that day you hope you run into everybody you know because you feel absolutely fantastic about yourself. Your spirit is soaring. Your self-esteem and self-confidence are high. You feel really good about yourself and your ability to interact with others is on much firmer ground. You don't become arrogant or self-important because of it—you become fully functioning. In fact, it's so dramatic because you leave your own personal hang-ups behind and become much more capable of sharing with and contributing to other people. Your effectiveness at your work and at human relations makes a sudden and dramatic improvement. It's all because of your hair and how you feel about it!

It's really a remarkable phenomenon. When your hair looks great you send out a very positive energy and people are naturally magnetized to it. They want to be with you, they want to be around you, they want to be like you. I always take pause at how hair has popularized public figures and been a distinguishing feature of attraction.

Now, before I continue, I want to emphasize that other aspects of appearance, such as skin, nails, and make-up, are also vitally important to how a person experiences life. I'm using hair in my discussion because it's something everyone, male and female, can readily relate to.

THE PSYCHOLOGY OF GROOMING

The choice and use of grooming products has a dynamic psychological impact on people. By choosing what is used, a person is making a meaningful statement about personal feelings of self-worth. Furthermore, the expectations and values that are linked to the use of specific grooming products carry powerful psychological energy. We have the opportunity to provide meaningful psychological service to our guests through the medium of the "home maintenance systems" we provide.

SELF-WORTH VALIDATION

Grooming products give people an opportunity to validate their feelings of self-worth. A product carries along with it a lot more than just what's inside the container. The packaging connotes particular imagery. The advertising focuses on specific values. The personalized directions and demonstration by the cosmetologist carry remarkable value to the client.

Spending more on personal grooming products to which much value has been attached is really an opportunity for self-validation. It's an opportunity to say "I'm worth it! I deserve the best! I matter. I take the very best care of myself." It's an opportunity for self-affirmation.

Interestingly, many people want to spend more. They want to obtain what they perceive is the very best because they're really making the statement to themselves that they're the very best. Some consumers will scoff at products because the items aren't high priced enough to carry the image of exclusivity. Many people insist on the best and believe that paying more guarantees them the best.

We have clients who could use a little shoring up of their self-images. Quality products with value attached to them can begin to move these folks in the direction of feeling better about themselves. Just the process of stepping up to pay a little bit more for their personal grooming products—to splurge on themselves—can have an enlivening psychological effect.

When you stop and think that people are using these products during very personal and intimate moments—and that they have a cleansing and rejuvenating effect—and that they're used in a ceremonial and consistent way—the result is an anointing and healing and blessing dynamic that operates psychologically. Take your own experience.

Imagine that you were away in the country for a weekend holiday. When you checked into your lodgings you discovered that you forgot to pack all your skin care products. The only thing available was the tiny bar of deodorant soap supplied by the motel. That's what you had to wash your face with. We've all found ourselves in this kind of predicament.

Now, as you're washing your face with that little bar of deodorant soap what thoughts are going through your mind? Not very positive ones, I'm sure. "Get it on and get it off quick." You may even be concerned that you're doing your skin a disservice! You may fear dryness, damage, and blemishes! There's no positive psychological charge out of that experience. If anything, it's negative.

Contrast that with the first time you "treated" yourself to a skin care regime for home use. Perhaps you were in the big department store and were taken by all the beautiful and glamorous skin care displays. One of the "consultants," as pleasant as can be, started talking to you about your skin and began recommending a specific regimen of skin care for home. She demonstrated the cleanser and toner and mask and scrub. She let you try some of the different creams and lotions and designed a whole program for you to use at home. You purchased several of the items.

Chances are that you pulled them out of the bag and started using them within a short time of arriving home. What a treat! At the very least you brought them into play the next morning. As you used the products what were your feelings? What were you thinking about?

Two thought patterns were predominant in your mind:

1. You did your best to make sure you were following the instructions with all the precision you could manage. You wanted to make sure you received the maximum value and benefit from your investment. You desired to have it all. You were serious about it.

2. As you were using the products you washed over your mind all the positive thoughts and images that you had attached to using the magic potions. Thoughts of youth, romance, glamour, sex appeal, prestige, recognition, and the like showered your mind. You felt good about yourself. You felt good about your prospects. You felt good about the world! You were enlivened!

Now you know about the potent power of the psychology of grooming. You've experienced it. It's real!

SALON PRODUCTS PROVIDE "AFFIRMATION IN A BOTTLE"

The same dynamic is operative, and perhaps heightened, with the grooming preparations we proffer at the salon. All the skin and body care balms, all the nail, hand and foot care liniments, and the hair and scalp care potions—they all can pack substantial psychological wallop!

Sometimes, because we're using these items every day we may get a bit complacent about them. To us they're normal. However, to clients they are a big treat! They love them, they love what they do, and they love how they feel when they use them.

It's rare to find clients inclined to visit the salon daily. Yes, they love how they feel when they leave but it's just not practical for us to see them each day. And we're certainly not in the position to go visit our favorite clients personally and do their hair each day. However, I discovered a long time ago that the way we can be with our clients each day is through the medium of the product. The product can be our representative, our energy in the household of the client.

What we say about the product, how we talk about it, the benefits and values that we link to the use of the product—by that process we literally infuse the product with a tremendous positive energy. It's as if we added another ingredient. It's as if we put a genie in the bottle. And the fact is that when the client uses our potion, all the positive energy we associated with the product comes right out of the container along with the other contents. Merely the fact of the container in the client's home radiates our positive energy (Fig. 4-3).

Call it "affirmation in a bottle!" It's psychological potency we deliver with every container because of the dynamic of the "psychology of grooming." When you realize that the potions are used daily, and in the morning when the subconscious mind is particularly open to positive programming, it's clear we have a remarkable opportunity to provide an ongoing psychological service to the client.

It requires effort to demonstrate and motivate clients to use our home maintenance products. I believe strongly that it's worth

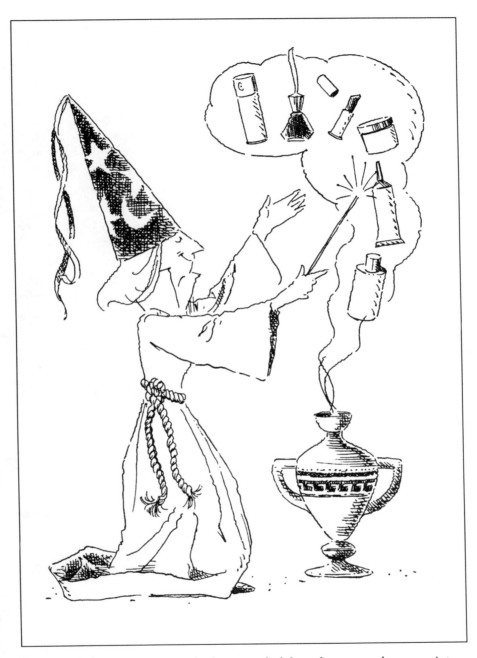

Figure 4-3   You make magic by how you link benefits to your home maintenance recommendations. That is why the value of what you prescribe is more
than simply what's in the bottle.

the effort. First, the value to the client is potentially enormous. The gift of daily positive energy is profound and can genuinely influence their experience of life. Furthermore, it's a matter of professionalism. To be aware that we could provide this service, and yet decide not to, is a sad state. It could only indicate that we're letting our own psychological roadblocks hold sway.

Occasionally designers complain that there isn't a lot of money in retailing. Well, there isn't a lot of money in doctors' writing prescriptions either. But they do it out of a sense of professional duty and in service to the client. My experience is quite clear that a direct link exists between client satisfaction, client retention, and client referral and getting clients on home maintenance systems. This only makes sense. With your bottles in their private quarters, they think of you more often, which holds a great deal of positive energy. The effects of retailing are so overwhelmingly positive from a professional and client service standpoint, that any commission or bonus earned is a really secondary concern.

COMMUNICATION STRATEGIES TO MAXIMIZE THE PSYCHOLOGICAL SERVICE

Quite naturally, the entire atmosphere of the salon, your own personality and design confidence, the work itself, and the client's own internal dynamics play a substantial role in the psychological impact of the created image. However, our "chair-side" manner, the specific images we convey, and the pictures we paint with words are of profound influence in delivering the psychological part of our product.

Three communication strategies are of substantial benefit in helping the client enjoy the full psychological bounty associated with our services and products. They are imbedded commands, psychologically charged expressions, and the practice of putting the client in the picture.

IMBEDDED COMMANDS

How you say something dramatically influences how it is understood. The field of neurolinguistics studies how people understand communication. One way to dramatically vitalize understanding and impact is through the use of imbedded commands, a neurolinguistic technique. The nice thing about imbedded commands is that they're very easy for us to use. You only have to remember one key word. It's a word you already use many times each day and the word is "YOU."

"You" is the word. Of course you can use a derivative of the word "you" like "you'll" or "your" or "you're" or even "y'all"! The root word "you" is the key. The strategy is to begin as many sentences, expressions, and phrases as possible with the word "you."

When you say "you" to people, they automatically sit up and pay attention because you're talking about their favorite topic, namely themselves! The information you share is personalized. They automatically relate it to themselves. And, if it's at all instructional or predictive in nature it has a conditioning effect. Expectations and images are created in the mind. Accompanying that, especially if the message is positive, is an automatic drive and desire to experience the expectation as reality. That's the command feature.

How to Use Imbedded Commands. The most fundamental place to install imbedded commands in your client communication is in your discussion of product or service features, benefits, and values. It's so easy to put the focus of attention on the product or technique and get into "it" discussions. "It does this . . . it does that . . . it will make this happen . . ." Actually, that's all quite bland and not nearly as fascinating to people as themselves. Furthermore, they don't automatically relate the information to themselves.

Use of imbedded commands ends all that. Instead of talking about "it," begin to use expressions like this:

You'll notice	You'll discover	You'll find
You'll appreciate	You'll enjoy	You'll experience
You'll receive	You'll love	You'll want
You'll apply	You'll massage	You'll feel
You'll see	You'll smell	You'll try

You get the idea. You understand how this works. You'll immediately notice your clients will respond to your suggestions more positively than ever before. You'll appreciate that your consultations have become more influential for you. Your confidence will grow, your effectiveness as a communicator will expand, and your clients will love you for it. You see?

PSYCHOLOGICALLY CHARGED EXPRESSIONS

Another method you'll want to master is the use of psychologically charged expressions. Keep in mind that the goal of salon guests is to receive what they want. They want to project the image that reflects their personality and creates the desired reaction in others.

As we are discussing services and products, it is crucial that we assure our guests that they will receive what they want. This is not so much in terms of how they will look as it is in terms of how they will feel about how they look and how others will feel about how they look. It's these feelings that they really desire.

In my seminars I often ask, "How do the salon's guests want to feel?" The most common response is "good." That's true, they want to feel good. The problem is that "good" has no psychological sparkle to it. "This perm will look good" falls flat. "The highlights will make you look good" just isn't enough of a call to action. "Good" doesn't build desire or motivate. So, we have to expand our vocabulary and talk to people where they live.

Salon guests are individuals and all are dissatisfied about the image they're projecting. The question is "why the dissatisfaction?" Is it because their hair, for example, makes them look sloppy, unprofessional, unkempt, slovenly, lazy, ordinary, and careless. If that's what's bothering them, then assure them that they'll receive what they want. They'll look polished, professional, well tailored, well put together, well detailed, executive, smart, and sharp.

Now you're meeting their psychological requirements. To take it a step further, if you assure them that particular services and products are the means to their psychological satisfaction, you'll stimulate countless impulse purchases. For example, talk to a 40-year-old slightly graying gentleman about haircoloring and he doesn't relate. Talk to him about youth, romance, sex appeal, business competitiveness, success, earning power, and masculinity and you have his undivided attention. Haircolor is merely the means by which we satisfy these psychologically based cravings.

So, to be most effective you'll want to put the major emphasis on psychological imaginings and deemphasize the technical means of delivery. You're the expert anyway. You can use the step-by-step technical summary as a way to justify price if you wish.

Fifty Words with Psychological Charge

safe	secure	assured	certain
guarantee	confident	young	youthful
sexy	romantic	desirable	appealing
passionate	handsome	executive	distinguished
wealthy	successful	prestigious	respected
polished	refined	glamorous	fashionable
admired	recognized	leadership	beautiful
cute	adorable	strong	virile
masculine	feminine	thin	slender
accepted	admired	appreciated	charming
attractive	popular	loved	liked
happy	positive	seductive	enticing
provocative	prosperous		

Use psychologically charged expressions abundantly. We have a very rich language, but here's a good starting point. In using psychologically charged expressions you simply link the satisfaction of the psychological desire of the guest with receiving the prescribed service or using the recommended product. "You'll immediately feel more confident and secure that you're projecting polished professional image that's well received when you use dependable XYZ spray." "You'll feel more desirable and alluring and suitors will find you sexy and romantic with this ravishing haircolor."

Feel comfortable reassuring clients with this kind of language. You'll find that they really respond to you. You'll notice that they're very responsive to the ideas that will deliver these results that they desire. You'll understand that it's an opportunity to give to your clients. Don't be bashful or shy. Don't think that you're being too personal. You'll discover that people will hang on your every word and suggestion.

A particularly strong opportunity exists with the female designer and the male client. Gentlemen rarely receive feedback about their appearance and how it can be improved to make them more appealing and attractive. Give it to them. Of course, suggest the array of treatments, coloring, and texturizing services to your male clients. Let them know clearly that they'll project the appeal and confidence that will make them irresistible when they have their image "designed."

PUT THE CLIENT IN THE PICTURE

Let people experience in advance that the services and products they obtain will lead to the culmination of their psychological desires. Let them visualize and be assured that they'll feel good about themselves and be well received by those important to them after their time with you. Paint word pictures. Put them in the future so they can encounter for themselves the reality they want so much. That's what "putting them in the picture" is all about.

However, not only does putting them in the picture have a reassuring aspect, it also dramatically builds desire for the benefit of creating products and services. Because of its power-ful "fortune telling" character, this strategy has the power to boost and uplift the client's actual experience of the future. This is a powerful tool because people automatically visualize the future.

Think back for a moment about the last time you pur-chased a wonderful dress or suit. Imagine the variety of clothes that you tried on and saw yourself in. As your selection narrowed and you were looking at yourself in the mirror, where did you see yourself? In the store? I think not. You saw yourself in the sit-uation where you were planning to wear that new outfit. A wed-ding, a reunion, a date—that's where you saw yourself.

As you looked in the mirror you asked yourself how you felt about the image the outfit projected. Did it make you look good to yourself? Did it make you look good to others? Did it reflect your personality and your sense of style? How would others respond to you? What sorts of looks would they have on their faces? What sorts of conversations would you be having? How would you feel about everything? See, you automatically put yourself in the picture. If the person in the clothing store was alert, he or she would be guiding you through the picture, pro-viding reassurance and "fortune telling" good all the way around.

The very same dynamic is operative with hair, skin, make-up and nails. People automatically put themselves in a future picture to experience their feelings and the reactions of others to their look. The more dramatic the color or style change, the more certain this is to happen. Naturally, we want them to experience positive feelings and carry themselves with con-fidence and self-assurance. So, if we're proactive about "putting them in the picture" we actually provide a vital psychological ser-vice. And we strengthen our relationship with the client at the same time.

Use All the Physical Senses. Make the picture vivid. Bring all the physical senses to bear—sight, sound, taste, smell, touch. The sense of sight is the most vivid and our services are appearance services so be sure to picture what is seen by the client and what the people with which they come in contact see. Relate what is said and what is heard. What are people thinking? What is everyone saying? What is being felt? What is being touched? How about emotional feelings? Is the breath quickening? Is the heart palpitating? Is passion coursing through the veins? Is the mind racing? Or is there a feeling of relaxation and quiet confidence?

When people undergo an image change, especially when it's daring, they often fear that they may be making a mistake. When we put them "in the picture" we cannot only neutralize those fears, but suggest a highly beneficial outcome and you'll help create the "head space" to bring it about!

Put Them in a Picture That Counts. To have the greatest impact, put your clients in a future picture that really matters to them. Try to find a setting that they already have a lot of hopes and expectations about (Fig. 4-4). What's happening in the client's life? Where are they going? Who are they seeing? Is there a special occasion that brought them to the salon to start with?

Everybody has something going on—a new romance; a new job or a promotion; a reunion, wedding, or family gathering; a vacation or convention; a big party or social gathering. Perhaps they've just lost weight or are starting on a self-improvement program. Maybe they're new in town. Maybe they've got a big competition coming up. Find out something of consequence and put them in that picture. Describe in vivid detail how they'll feel about themselves and how others will feel about them in that setting they care so very much about.

COMBINING THE METHODS

Your most potent psychological energy comes when you integrate all three methods of imbedded commands, psychologically charged expressions and putting them in the picture. It makes your recommendations and prescriptions irresistible and creates a long positive afterglow to the visit.

Here's an example of John who's just starting an important new job on Monday and is in for a haircut the weekend before.

Figure 4-4 Help your clients visualize the benefits they want by putting them in the picture and you'll improve their confidence and self-assurance.

We'll use these techniques as we're styling his hair and demonstrating our favorite home maintenance potions.

John, before you blow dry your hair and go into the
office on Monday put a little XYZ brand sculpting
lotion in your hand like this and distribute it evenly in
your hair like this. Notice how light and fresh it is on
your hair. Now you hold the blow dryer like this and the
brush like this and look how your hair turns out so full
and polished. You get a real professional and executive
look—very well groomed and detailed. You finish it off
with a little XYZ spray that makes your hair look so
healthy and shiny and you've got that A-1 'dress for
success' image. Now when you walk in the office on
Monday, perk your ears because you're going to hear
whispers from the secretarial pool. They're going to be
saying 'who's that hunky young executive who just
was hired?' John, you'll have your pick of the crop! And
when your boss lays eyes on you, he'll feel so confident
he'll immediately say to himself that you present a perfect image for the company. You'll be on the fast track
immediately! And when you interact with clients you'll
feel so comfortable and confident. Your self-assurance
will radiate and others will respond and take you seriously in a friendly way.

Now John, on the weekend when you're out socializing, you'll want to look more contemporary and stylish, so take a generous dollop of the XYZ brand styling
gel . . . like this . . . and work it in your hair . . . like
this . . . and you'll notice the definition and raw mas-
culine appeal your hair takes on. You walk into one of
the popular nightspots and you'll have the young
ladies all giving you the once over. You'll be the hottest
dude on the town. And you get out there and boogie all
you want because you can have the confidence that the
gel will give your hair the holding power to keep your
style looking sexy and alluring all night long. So, when
you want to make dates and exchange phone numbers

at the midnight hour, you have the self-assurance of knowing that you look just as handsome and desirable as when you first walked in the door!

Wow! Who could say no to that? Notice how all the elements were integrated into the conversation. This is what having a good "chair-side manner" is all about! This is how you keep the conversation focused on hair and on the client. When you're talking about the client you'll have the person's undivided attention. Clients will be intensely interested in what you have to say and demonstrate for their beneficial use. This is bonding with your clients. This is exactly what they want you to be talking about and what they want you to do for them. And, it's so easy to say a few positive words while they're sitting there in the chair—all we have to do is throw a little air over our voice box and wiggle our tongue a little bit and that's hardly any effort at all!

Whether you bring all the elements together for a color, perm, relaxer, cut, or style, the same psychological dynamic is operative. How you talk about new make-up, better skin care, or more exciting nails gives you the opportunity to create the same psychological feelings. And when the client is in his future moments, your words become the stuff of strength.

When John is getting ready Monday morning to go to the office he's thinking about all the ideas you implanted. When he's getting ready to out on Saturday night he's visualizing the expectations you implanted. And when he's on the firing line Monday and Saturday, his ideas, thoughts, beliefs, attitudes, and behaviors are influenced by the anointing power of the magic potions you've provided for him.

That's why the psychology of grooming is so valid. The products are the means and medium of conveying the message. Without the product the psychological feeling simply isn't as intense—if it can exist at all. The medium of the product is essential for the psychological dynamic to bear fruit. The client's experience of the events can be altered by the psychological seeds you sow in advance. This is when our influence has a long afterglow and our impact on the client is meaningfully heightened.

SUMMARY

Your guests want to feel good about themselves. The psychological reassurance you provide is of monumental importance. In this chapter we discovered:

- Salon clients come in dissatisfied that they're not projecting the image that reflects their personality.

- Creating and communicating a satisfying image for the guest is an art and can positively affect the client's quality of life.

- The home maintenance systems we provide can let your client's psychological well-being linger and be reaffirmed daily.

- Our communication using imbedded commands, psychologically charged expressions, and word pictures is beneficial and enlivening for salon guests.

It's worth the effort to massage the psyche of salon visitors. Your service ends up having considerably more value. Truly, the more you help a person experience the heart's desire, the closer you come to experiencing your own heart's desire.

CHAPTER 5

The Psychology of Success in Cosmetology

66 ——————————————

Success is not something you obtain; success is something you prepare yourself to receive.

——————————————— **99**

To enjoy success, you have to believe in the fundamental merit of what you're doing; you have to believe in your capability to do it; and you have to experience a level of happiness, joy, and fulfillment in its pursuit. These are all essentially psychological states. The good news is that we have the power to control our thinking. The challenge is that we are often enslaved by old patterns of thinking that quash the desired successful mind-set we want.

That successful state of mind is necessary and essential if we are to enjoy a fruitful life and career. Indeed, there can be no success without it because a positive mental life is perhaps the most readily discerned characteristic of success. Make no mistake, our psychological state is pivotal in determining how we actually experience life. Remember, the mind on its own can make a heaven of hell or a hell of heaven.

WHAT YOU WILL DISCOVER IN THIS CHAPTER

- You will learn about self-image psychology and how your beliefs about yourself were formed.

- You'll discover how your self-esteem affects on your performance and how fine-tuning your feelings about yourself can propel you to new horizons of growth and excellence.

- You'll find out how to rethink your circumstances and see yourself in a brighter light than ever before.

- You'll learn how to take charge of your mental life and direct your thinking in healthy and positive ways.

- You'll become acquainted with the universal law of success and find out how contribution is the key to happiness and prosperity.

DEFINITION OF SELF-IMAGE

I once saw on a billboard that "the me I see is the me I'll be." How do you see yourself? What kind of a person do you think you are, really? How do you feel about the value of your professional career?

Each of us has a self-image or self-concept. It is the bundle of things we believe about ourselves. We have a concept of ourselves in all the different roles that we play. This concept reflects our beliefs about how "able" we are in the different situations we encounter. No doubt, in some situations you feel masterful and able. Likewise, in other areas of your life you don't feel very proficient. So, our self-concept has to do with how we define ourselves—who we believe we are—and just as importantly—what we believe we're not.

Naturally, self-concept is involved in every aspect of our career. There are some things we think we're good at and some things we think we're not so good at. This is true no matter what aspect of the industry you're involved in. Designers may feel they execute some cuts well and fall flat on others. People may have certain colors they favor because they're confident about their ability with them, and then shy away from others altogether. Owners or managers may think they're good at team management but poor at financial management. They may believe that they can put together beautiful displays but then think they have no luck with advertising campaigns. Salon consultants may think they work well with chains but not so well with independent contractors—or vice versa.

THE ROOTS OF OUR SALON SELF-CONCEPT

Our self-concept is rooted in our experiences. How we performed a given task in the past, and the reaction to that performance, influences how proficient we believe we are. As the great writer Oscar Wilde said: "Experience is the name so many people give to their mistakes." Whenever we're trying something new we're going to make a lot of mistakes. As a matter of fact, an expert has been defined as the person who has made all the possible mistakes!

So the chances are that as you were learning your salon skills you made some mistakes. Do not be surprised! It's also likely that with some things you got it perfectly right from the start. You take a list of a hundred salon procedures and try them all for the first time and odds are that you're going to do some of them well and some others not so well.

Often, it's not the procedure itself, it's just the law of averages. When it's a more complicated or sophisticated task, the law of averages is going to allow a smaller percentage of "first trys" to succeed! This is just reality. It's nothing personal. In fact, some things are so challenging that it may take a dozen trys to get it right. Using an objective standard, we know that we can't take the natural learning curve too much to heart.

However, it's the reaction to those early experiences that cement your self-image in each area of activity. Reactions vary. How you were feeling on the given day, the setting, the presence of others, your expectations, the expectations of others, the real or imagined consequences, the words that were spoken silently to yourself, the words that were said by others—all these and more make up the reality of the reaction.

When the reaction is positive, we have self-confidence about our ability to perform the particular function. When the reaction is not positive, we can assume a variety of attitudes. Maybe we think "better luck next time" or "practice makes perfect." That's a healthy approach.

THE CREATION OF NEGATIVE SELF-CONCEPTS

Depending on the intensity and nature of the reaction to our performances we can jump to all sorts of wild conclusions like "I'm no good at this" or "I'll never be able to do this" or "I'll never try this again." Wow! Those are big jumps. Logically, when it comes to common procedures that are performed well by thousands of colleagues every day, we know that these sweeping negative responses simply don't make sense. When a load of emotion is infused into a situation, reactions can become deeply rooted.

Let's say for example you tried a new perm wrap technique and didn't get it quite right. If the client was half-way satisfied, you think "I'll get it better next time." But let's say a second client made a lot of noise and got the salon owner involved and refused to pay and had her husband call in the evening to threaten a law suit! Now we have a stronger reaction. You may not be so eager to try that perm wrap ever again. But keep this in mind, the actual results on the two clients could have been exactly the same. It's simply the reaction that intensifies your self-concept of performance positively or negatively.

Let me give you another example. Let's say when you were in school and working on a mannequin head you made a mistake on a cut and the instructor berated you about it and made you feel and look foolish in front of your classmates. You'd probably feel awkward about attempting that cut for some time because so much emotion was infused into the situation.

Then the very next day you had a live client on whom you were doing an entirely different haircut style and it just wasn't working out right at all. A supportive instructor came to your side with words of encouragement. This person took the time to show you the way through the cut and made you feel like you were making positive progress every step of the way. I'll bet you'd want to try that cut again as soon as possible because you'd want to re-experience the positive energy associated with it. It's these kinds of incidents that root our self-concept.

You have a self-concept about everything, and, make no mistake, it goes beyond your life as a salon professional. Your self-image lives in every role you play—as a wife or husband, as a son or daughter, as a citizen, scholar, athlete—the list is endless. Your self-image defines who you think you are, what you think you can do, how you believe you will perform. It also defines the flip side of the coin—the self-image reflects who you think you are not and what you think you cannot do.

Actually, your self-image ends up being a lot like a self-fulfilling prophecy. Generally, what you think you're going to do well with, you'll do well with. What you think you'll do poorly with, you'll do poorly with. As you believe, so is it. It's a teaching as ancient as the scriptures. The ray of hope is that you have the power to choose what you're going to believe about yourself and you can start the process of adjusting your beliefs any time you choose to.

YOUR SELF-ESTEEM

Your self-esteem gets down to the issue of how you value yourself. It flows from your self-image. Your self-image, or self-concept, tells you who you've come to believe you are in all the different areas of your life. Your self-esteem then evaluates those descriptions and judges them good, bad, or indifferent.

We live in a culture that's rife with low self-esteem. I suffer from it in some areas of my life and so does everyone else I know. We work in an industry that has a self-esteem problem, where many cosmetologists are almost ashamed to proclaim their vocation with pride and a sense of high value. Couple that with all the negative personal baggage we lug around and you've got a lot standing in the way of a successful psychological state of mind!

As humans, we so often focus on what we don't have rather than on what we do have—on what we can't do rather than on what we can do—on our lack rather than on our abundance—on our weakness rather than on our strength.

Also, we often tend to evaluate ourselves very harshly. We are usually our own worst critics. We often feel that we don't measure up. We feel that we're not good enough, that somehow we are less than others. Success-oriented people, especially, judge themselves mercilessly. I've often said it to others that if anyone else dared to talk to me the way I sometimes talk to myself, I'd be furious! Yet so often we give ourselves a steady diet of negative messages.

It's actually remarkable how the mind has a tendency to define and focus on things negatively. People who have everything in the world going for them by any objective standard can develop an outlook of misery and woe and self-pity. It's actually quite absurd.

CHANGE YOUR PERSPECTIVE AND YOU'LL CHANGE YOUR SELF-ESTEEM

When we have so much going for ourselves and things are 90% great, we'll often end up focusing on the 10% that's causing us some angst. The challenge comes down to how we're looking at and defining that 10%. If we see it as a problem, then it's a problem. If we see it as an opportunity for growth and development and personal adventure, then that's what it becomes.

Now you can begin to see how the self-concept and self-esteem play off of each other. We first come to believe that we're a certain way. We follow that up by evaluating that definition. Then, depending on our perspective and approach, our self-esteem can be hopeful or hopeless. If hopeful, we accept ourselves and give ourselves room and time to grow. That's positive. If hopeless, we engage in a constant barrage of putting ourselves down in a way that no one else would dare! It becomes particularly pitiful if we get into the destructive loop of "I have to, but I can't—I have to, but I can't." Unfortunately, an inadequate self-concept and counterproductive self-esteem amounts to a big ball and chain of fear and psychological enslavement holding us back from growth. The bottom line is that the psychological state that success demands requires a healthy, positive self-esteem.

IMPROVED SELF-ESTEEM ENHANCES RELATIONSHIPS

As we improve our own self-esteem, so too our feelings about others undergoes transformation. On reflection you'll discover how often our critical and negative thoughts about others is rooted in our own low self-esteem. A symptom of low self-esteem is the need to act superior to and demean others. People who do this feel so poorly about themselves that the only way they know how to shore themselves up is to compare themselves to others. The process of finding flaws in others, which is essentially a negative exercise, helps distract them from having to look at the character defects in themselves.

Individuals with truly high levels of self-esteem don't need to belittle others as a transparent effort to build themselves up. If anything, those with high self-esteem have an understanding of human life that gives them a compassionate and gentle view of others. They support this with encouraging and kindly behavior. As the rule of Ancient Greece put it so poetically—"as within, so without, as above, so below."

THE PROCESS OF CHANGE AND GROWTH

It's not so much whether you "believe in yourself"—but rather what you believe about yourself that's holding you back. If what you believe about yourself is not getting you the life and career you want, then you must start by changing what you believe about yourself. That alone starts the process of real and lasting change. Let's review a number of powerful ideas that can alter how we look at and feel about ourselves and the world we live in.

UNHOOK FROM NEGATIVE IMAGES OF THE PAST

Yesterday is history. Your past is gone. There's nothing that can be done to change it. The only thing that can change is your attitude toward it. The first stage of growth is to accept your past and bring it to closure. You want to be able to unhook from the patterns of the past so you can reinvent and redefine yourself as the person you really want to be.

Some of us were told by parents and other adults in authority that we were not acceptable in one way or another. Some children were subjected to dysfunctional family behavior patterns and unloving ways of communicating and relating. Rich or poor, urban or rural, white collar or blue collar—practically everyone has some tales from youth that continue to haunt them into adulthood.

Fault finding, blaming, anger, resentment, self-pity, and bitterness are not positive states of mind. Engaging in this type of thinking is poisonous to your mind, body, and spirit. Too bad that some important adults in our lives misbehaved or never learned the skills of reflecting love, understanding, and compassion. But, it's in our hands to break the pattern. We have to learn to accept their limitations and understand that if they had the ability and presence of mind to relate more lovingly, they surely would have. The fact is that they played with the best cards they had. A spirit of understanding and forgiveness on our part is the only positive direction in which to move. Truly, it's an act of self-love!

There's an old Zen philosophy that teaches that when you return right with right, good comes about. And that when you return wrong with right, good also comes about. Let early injustices in your life be brought to closure with a spirit of goodness. Use these circumstances as an opportunity to exercise your highest virtue and character. At all cost avoid the mistake of letting past misfortune become a justification for current stagnation and failure in life. That's much too high a price for you to pay today—and you don't have to pay it! Your life has unfolded as it has. It's really provided opportunity to learn the poignant lessons of loving and giving. You have the power right now to transform any negative feelings about your past into positive opportunities for personal growth.

UNHOOK FROM NEGATIVE HABIT PATTERNS OF THE PAST

Most of us have developed a host of counterproductive habit patterns. Sometimes they are thought of as the seven deadly sins: pride, covetousness, lust, anger, gluttony, envy, and sloth. Interesting to note that these are all states of mind and patterns of thinking that lead naturally to obsessions and behaviors that are neither healthy nor loving. I don't know anyone on earth today personally who isn't wrestling with one or more of these bugaboos. After all, we're human!

On reflection, you'll discover that the root of these recurring issues is our own fear. At one time or another in our life, fears that we were not going to get what we wanted or fears that we were going to lose what we had, generated compulsive, knee-jerk responses that are unworkable. Fear-based responses usually emanate from the darker side of human nature.

If the same unfortunate personal and behavior problems keep recurring in our lives, it's because we're stuck in a fear-based response mode. As the American poet Edna St. Vincent

Millay remarked "it's not one thing after another, it's the same thing over and over." Simply looking at the unmanageable parts of your life will show you where the defective habit patterns are at play. Situations and relationships that bring up intense feelings of anger and resentment also clue you into unworkable habit patterns in your life.

First and foremost, learn to recognize the patterns. Understanding that the only thing that you can change is yourself, you'll want to understand the role that you play in these areas. Put the focus on yourself and distilling what you may have thought or said or how you may have behaved that led to the situation. Because there's nothing we can really do to change others or bring others under our control, the energy expended in shifting blame and finding fault in others is futile and will only magnify the pain of the situation. Sadly, it's not unusual for a person to get an almost perverse pleasure out of wallowing in the darkness.

When you discover an unworkable pattern in your life, make a commitment to try a new response the next time. Make it a response filled with light and love. I find that reading inspirational literature and spending time in contemplation enables me to receive insight on a positive response to work with. Over time and with a little experience in working with the new behavior, it's amazing how we can grow and transform into more fully functioning and capable human beings.

Accept Yourself as Being OK. We're not perfect. If you can make measurable progress in reasonable time then you're definitely moving in the right direction. The need to be perfect and have all your circumstances reflect perfection is called "perfectionism." It's worthwhile to understand that the psychological community considers "perfectionism" to be a personality disorder. Our circumstances, relationships, and behaviors aren't perfect. Thinking that we can control people, places, and things—and even ourselves—to the point of perfection is going to result in frustration and anxiety. We have to learn to accept ourselves as we are—both good and bad—and do our best today.

Fat or thin, tall or short, female or male, rich or poor—whoever you are, you're OK. You are a child of the universe. You are a creation of love. You are fine just the way you are. It's OK for you to be exactly who you are. You have no excuses to make to anyone. Plant your feet firmly on the ground and say "this is me." Be yourself! You never have to pretend to be somebody or something that you're not because you have all the potential for greatness and magnificence in the world just as you are.

One of the great causes of tension in life is the failure to accept ourselves. If we're unhappy with our circumstances or relationships, we can slip into feelings of anger or self-pity. Anger is a natural response to injustice. However, when we're not completely satisfied with our own behavior—where we feel out of control or unable to control—we can turn the anger inward against ourselves. This is not positive. Self-hatred doesn't nurture self-esteem, as you can imagine.

Accept the World. The other bugaboo pulling at our mental equilibrium is self-pity. Self-pity is the self-centered emotion we feel when we start thinking that the world and its people aren't treating us right. When we feel victimized and not in control of events, anger and bitterness can follow. Saying that the world is not right, or fair, or just is meaningless. It isn't going to change anything. That kind of thinking leads nowhere fast—it's a dead end street. The world is just the way it is.

A reality we must accept is that the world is not going to change to accommodate us. The world owes us nothing. We must change to accommodate the world because we cannot change the world. We have to work on the only thing that we can change—namely ourselves.

The world is interested in our contribution. The soil of opportunity is waiting for us to till, plant, fertilize, and harvest. The earth doesn't say "bring me your need"; it says "bring me your seed!" As night follows day we will get from the world exactly what we deserve. If we don't deserve much we won't get much. The word "deserve" itself means "from service." We must accept that we are the cause of the results we experience. If we want our results and circumstances to change, then the simple truth is that we have to change. We have to refine and improve ourselves to get more in tune with the world the way it really is.

REDEFINE YOURSELF POSITIVELY

Before you can go anywhere else, you must plant your feet firmly where you are right now. It's OK for you to be there! Take a positive look at where you are right now. Focus on all the good things about yourself and turn your attention squarely to your assets and away from your liabilities. We possess a magic magnifying mind that amplifies whatever it tunes into. That's why it's so critical for us to learn to focus on our strengths.

Take a piece of paper and list 10 things that you've accomplished in life that are a source of pride. Then ask yourself, "what does accomplishing these 10 things say about me?" To achieve these results means that you've got some highly admirable character traits. Your accomplishments are evidence of certain positive traits that have proven to be capable of creating positive results. Recapture that for yourself. Acknowledge that you have accomplished feats of merit.

I really want to encourage you to do this exercise before going on. It can be a step in the right direction of accepting all that you are that's good. Those qualities, those positive traits that you've demonstrated are things worth focusing on. To have a healthy mental outlook means that we open our eyes to see the good. I'll bet that in doing this exercise you'll remind yourself of wonderful realities about yourself that you rarely contemplate. So often we spend so much time dwelling on the negative over and over again while at the same time undervaluing the tremendous positive we possess.

TOP 10 ACCOMPLISHMENTS

What Character Strength Does
This Accomplishment Affirm?

1 _____   1 _____
                                _____
2 _____   2 _____
                                _____
3 _____   3 _____
                                _____
4 _____   4 _____
                                _____
5 _____   5 _____
                                _____
6 _____   6 _____
                                _____
7 _____   7 _____
                                _____
8 _____   8 _____
                                _____
9 _____   9 _____
                                _____
10 _____   10 _____
                                _____

RUN WITH YOUR WINNERS

Taking note of the positive traits you've already seen in your life
is going to give you a big clue as to what you want to develop and
nurture. The most successful people get that way not by being
good at many things but rather by being outstanding at one

thing. Take whatever ability you have, whatever comes to you naturally and run with it! Play your strongest card. Success is about developing excellence at one thing. Value yourself enough to give yourself full permission to go for it with gusto. As they say in the stock market "cut your losses short and play your winners long!"

In a very practical sense, you have a good self-concept and high self-esteem about activities that have already come naturally to you. Certainly, the beauty business can be a source of tremendous self-esteem. Chances are that there has already been a love affair between you and the image industry. I'll bet you've had some peak experiences associated with cosmetology. Some of your richest moments—some of your greatest pleasures—some of your fondest hopes and dreams are intimately tied into your career in cosmetology.

It could be the happiness you've enjoyed associated with performing a technical service. It could be the pride associated with completing cosmetology school or advanced salon education. It could be the joy you've received from designing and creating fantastic looks for your clients. It could be the happiness that has accompanied heartfelt words of thanks and appreciation from salon guests. It could be the recognition from winning the retailing contest. It could be any one of a thousand different things you've experienced in the salon industry.

If you've had experience in the salon industry that's been positive, then by all means give yourself permission to take it and run with it—to multiply your success and accomplishment—to allow yourself to grow in that area and be all that you can be–to be more as a human being so that you can give more to others— and enjoy the abundant harvest that inevitably comes.

Sometimes we hold back because we haven't made the affirmative decision to make our career more than just a job. If people have the idea that they'll work in the salon business until something better comes along, then their success will be automatically limited. If others have the attitude that they're not sure if working in the salon business is what they really want to do, then their success will be automatically limited. If they have a problem with accepting their role as a salon professional then their success will be automatically limited.

Why limit yourself? Play your winners full out! If it's worth doing to begin with, especially when it comes to your career, it's worth doing wholeheartedly! Your most precious resource is time. If you're going to invest your time then you're exchanging the most valuable thing that you have to give. You owe it to yourself to pursue your career with more than half-hearted effort.

PRACTICE PRESENCE OF MIND

Change Your Thinking, Change Your Life. You must take charge of your mental life. One of the few things in life that you can absolutely control is your thinking. Each thought is an act of choice. To experience the full measure of abundance, prosperity, happiness, and joy that is already yours, you must develop the presence of mind and the mental discipline to constantly focus your thoughts toward the sunshine and away from the shadows.

It's been said that if you change your thinking you can change your life. The human mind magnifies whatever it focuses on. If we focus on the problems we encounter then they dominate. If we focus on the progress we're making, then we add fuel to the engine of momentum. And the bottom line reality is that we have a choice regarding what we're going to focus on.

But there's a steady stream of thought that never ends. The self-talk and background noise of the mind goes on ad infinitum and almost unconsciously. Mental activity can run the full gambit from past to present to future. It takes a measure of awareness for us to pay attention to our active mental life and a measure of mental toughness to keep our thinking moving in the direction that's going to nurture self-esteem.

Your Attitude Influences Your Thinking. How you think about things will be largely determined by your attitude. Your attitude is the perspective you bring to situations and events. Focus on the good in things, see the silver lining, look for the positive in every situation and you have what is commonly called a "positive attitude." Focus on the problems, the difficulties, the obstacles, and the flaws and you have what is commonly referred to as a "negative attitude." If you tend toward a positive attitude, then your emotional response to situations will be more sunny and your disposition will be more magnanimous. If you tend toward a negative attitude, then your emotional response to situations will be cloudier and your disposition will be more discouraging.

The poet Rudyard Kipling advised in his distinguished poem *If*, that we "meet with triumph or disaster and treat those two impostors just the same." It's easy to be confused by the turn of events in life. We don't know how things are going to turn out and the fact is that situations run from hot to cold very quickly. When we build up great expectations about how we want things to turn out, we're setting ourselves up for frustration and unhappiness. Life can be so peculiar sometimes that what may seem to be positive today may turn out to be negative and what seemed like a negative turn today may wind up being a blessing in disguise. You know how it works!

The fact is that all we can do is what's in front of us right now and do it to the best of our ability; that requires a positive mental attitude. We have to learn to leave the outcomes to Mother Nature and just proceed with a sense of trust and happiness. Sage advice when you consider that most of us, when we meet a little turbulence, start casting our eyes and emotions toward the dark clouds. It's at those times that we can benefit from an "attitude adjustment" and recognize that our fortune-telling powers are pretty weak.

Your Attitude Is a Matter of Choice. No matter how up-beat and optimistic you are it's important to come to grips with the fact that from time to time situations out of your control can shake you into a negative outlook. This is just part of reality. Bad news, severe disappointment, unexpected accidents, illness, and loss can send anyone into a tailspin. The trick is to remain objective while allowing ourselves to experience the natural emotions of the moment. But then, as soon as possible, we should shift gears and move our attitude and emotions back to a positive frame of mind.

Let me give you an example. Let's say that spiral perm appointment that you'd planned the whole afternoon around calls in 15 minutes before show time and cancels; she doesn't even want to rebook it. Now naturally, you'll have a moment or two of disappointment when you come to grips with the loss. But then you have a decision to make. Are you going to wallow in the bad news or look for a rainbow?

On the one hand you can think of all the times you've had no-shows and last-minute cancellations. You can begin to question yourself. You can begin to question your career and profession. You can start indulging in self-pity. Poor me; my lot in life stinks. Life is unfair. And you can proceed to experience resentment against the salon owner or manager and coworkers. You can concoct reasons why they're to blame. You can begin questioning your whole career and the beauty business itself. You can remain in the tailspin for the whole afternoon if you wish and even carry it into the next day. You can make a life of it, and some people choose to.

On the other hand, you can choose to experience other thoughts that can move you in a positive direction. You can contemplate what you might have said or done that could have avoided this situation to begin with. Maybe you establish some new practices that can minimize these events in the future while not scaring clients off at the same time. Then you think about what positive you can do to invest this new-found time on your schedule. Perhaps you'll get an opportunity or two with a walk-in, and who knows, perhaps you can motivate one of them to experience the benefits of a perm today. Maybe there are some follow-up telephone calls that you now have time for or you can get to some thank-you notes or other correspondence. Perhaps you can set up a cross-promotion or begin putting together that newsletter that's been on your to-do list.

The bottom line is that you can either let the bad news send you into a spin-out, or you can develop the self-discipline and presence of mind to leave the negative on the side of the road and quickly proceed down the path of prosperity. The reality is that it's your decision.

ADJUSTING YOUR ATTITUDE

Sometimes we have to work at "adjusting" our attitude. Some people will take a walk around the block. Others will read or listen to something positive and motivational to remind them of the reality of the situation. Others might listen to some inspirational music to help change their mood. Another person will review goals and affirmations to get refocused. You may walk into the bathroom, look at yourself in the mirror, and give yourself a pep-talk. The key is to find out what works for you and then use it quickly when you notice your attitude needs adjusting.

Self-honesty is also important here. It can be easy to fall into the trap of saying "this time I'm really justified in being angry and resentful." The one you're hurting the most when you're angry and resentful is yourself. Look at it this way—it's like drinking a glass of poison and waiting for the other person to drop dead. You have to be realistic enough to know that your attitude is too precious a commodity to squander.

BENEFITS OF A POSITIVE MENTAL ATTITUDE

A positive attitude is particularly important in your career. Psychologists have studied the impact of attitude on all sorts of things and their conclusions simply make common sense. Here are four clear benefits you can look forward to with your positive mental approach.

1. *Your health and energy level is improved.* How you physically feel is effected by your attitude. Negative attitudes tend to attract illness and magnify discomfort that can result in absenteeism, tardiness, leaving work early, and an inordinate number of visits to the doctor. All of this has a dramatic impact on income in a way that isn't going to make your attitude any better.

2. *Your work performance is enhanced.* One thing is for sure, a positive attitude will make you more creative in your artistry. That will multiply itself and build your self-confidence

and customer satisfaction. It will give your self-esteem a boost and make you feel even better about the merit and value of your work.

3. *Your personal magnetism grows.* A positive attitude will also add sparkle to your personality and that's naturally attractive to people. A cheerful outlook will also make you appear more attractive. Isn't it remarkable? Even if a person is not "beautiful" by physical standards, she can be regarded as beautiful by virtue of a positive personality. And it's all because people want to be around you because they're magnetized by the appealing energy you send out.

4. *Your salon environment is "happening."* A positive attitude certainly improves the salon energy level. It helps elevate the spirits of both coworkers and clients. Others are really depending on us to have a positive attitude. The salon manager, for one, is counting on it. Team spirit—customer service—are attitude related. Some colleagues and salon guests have extremely difficult private lives. But at the salon there's an opportunity to be uplifted. We just simply gain so much additional fulfillment from the knowledge that we're providing healthy, positive attitude leadership. At the end of the day you'll feel that you've done something worthwhile. The fact of the matter is that we're spending a good percentage of our waking hours in the salon and we might just as well enjoy it and have a good time.

DEVELOPING A SENSE OF PERSPECTIVE

CULTIVATE HUMILITY

Humility is the profound awareness that we are not the center of the universe, that the whole world does not revolve around us. Shakespeare wrote how we are all merely actors in the great drama of life. Some of us have big roles; some of us have small roles. All the roles are really rather small in the context of the greater scheme of things. So we approach our work with a sense of humility.

The opposite of humility is self-centeredness. Self-centerness places us at the center of things. It's where we interpret events and circumstances only in terms of their impact on us as individuals. It's the mind-set that believes that our own wants and desires are the only things that matter and that other people and things are useful only to the extent that they can be used to get what we want.

With humility, we offer ourselves and our efforts to the greater good of all. We have a profound understanding that the world and it's people are not here to serve us, but rather we are here to serve our fellow man and whatever other purposes destiny has in store for us.

Think of the difference humility can make in our working with clients. If we're self-centered, we're primarily concerned with lining our own pockets and using clients to send us referrals. As a consequence, we may do excellent work, but our intentions do not reflect the highest and most admirable virtues. If we come from humility, then the clients are center stage. Helping them look better and feel better about themselves is the purpose. If we can play a positive role in the lives of our clients to improve their quality of life, then we're more in tune with the contribution we can make in the greater scheme of things.

Humility has nothing to do with humiliating ourselves. It has everything to do with understanding the true nature of our roles and accepting the outcome that nature has in store. Remember that virtue is it's own reward. We're talking about a philosophy of life where service to a greater purpose than ourselves comes first. Service to our fellow man, to our communities, to our nation, to our world, to our God.

CULTIVATE GRATITUDE

Gratitude is another element of equilibrium that ought not be underestimated. You've heard the importance of "counting your blessings." When we stop and think how good we've really got it we experience an immediate shift in attitude. The fact is, we have much to be grateful for.

Stop and think how well nature has provided for us. When was the last time you went to bed hungry? When was the last time you had to go without shoes? And you know the old story about the person who complained that he had no shoes until he saw someone who had no feet.

We manage to get what we need to survive. However, we don't always get everything we want. Nobody does. Without an attitude of gratitude there's a tendency for us to focus on what we don't have rather than be grateful for what we do have. When you cultivate an attitude of thankfulness and appreciation for all the good things you're enjoying in life, it's amazing how life seems so much brighter and joyous.

ENACT THE UNIVERSAL PHILOSOPHY OF SUCCESS

It's better to give than to receive. Focusing your energy outwardly and emphasizing what you can do in the service of others creates a greater feeling of satisfaction and mental well-being than anything else. One of the characteristics of purpose is that it's always outwardly focused. The great thinkers and philosophers of the ages repeatedly tell us that our purpose has nothing to do with accumulating or being on the receiving end. Rather, it has everything to do with giving and serving. The great paradox is that the greater the magnitude of our giving, the greater our capacity to receive. Receiving is a consequence of giving. Then the cycle continues with ever more giving followed by ever more receiving and then ever more giving again.

Many people are confused when they first face the universal philosophy of success. Sometimes it's easy to get more preoccupied with the idea of receiving rather than giving. A lot of this is due to the fear that if we give, we may not receive. Sometimes we become afraid that we'll somehow get shortchanged. Many of us were taught that we needed to get what we could while the getting was good, and that we needed to think of ourselves and our own wants first. Life was about winners and losers and beating others to the punch and making sure we got ours.

It's easy to understand how a person could fall into these ways of thinking. But in the final analysis they don't work. They assume the dark side of human nature. They put the focus on fear and guile rather than faith and trust.

It's interesting to note that one of the characteristics of great leaders is the confidence their people have in them. Great leaders have the faith and trust of their people because they've cultivated the image and reputation of being people of good character and virtue. A person of good character and virtue does what is right, fair, and just. The test of the strength of their character is that they'll do the right thing even when it's not in their own self-interest to do so. They'll use higher principles as guiding forces rather than the expediency of the moment.

Naturally, if we're to win the trust and confidence of clients, coworkers, staff, and business associates we must demonstrate virtue and character of the highest order. That means putting principles first and holding ourselves to a high standard of right action in all our interactions.

If the results we want are embodied in a sense of happiness, joy, prosperity, understanding, kindliness, and abundance then we must focus on doing that which will cause those results. It's all about service to others, giving the best of ourselves, helping our fellow man, and living our purpose.

Over and over Mother Nature has taught us that as we plant so shall we reap. As we give so shall we receive. The more seed we sow, the greater our harvest. The earth will not yield a harvest unless we give our seed care. Many of us have heard the parable of sowing and reaping. We sow our seed. Some is blown away by the wind. Some will fall on the rock and not take root. Some will take root but be choked off by the weeds. But some will land on fertile ground and grow. And the seed that grows yields a most abundant harvest—thirty-, sixty- or even one hundredfold from what we started with.

Every act of giving may not appear to yield direct results. But we give anyway, and we give with no attachment to the results. We leave the harvest up to Mother Nature and proceed with faith that the harvest will come as it always has. We often don't know where or how the harvest is going to come. And Mother Nature has an infinite number of ways to deliver our harvest. If we take the approach that we will only accept our harvest in one way and from one source then we are trying to limit the miraculous majesty of how Mother Nature is capable of yielding her bounty. And that seems to be a mistake because it usually winds up limiting how much seed we will sow and where we will sow it and, consequently, how much of a harvest we'll ultimately enjoy.

Perhaps we grew up in an environment where we were taught to approach life with fear rather than faith. Perhaps we were more exposed to the ideas of lack rather than abundance. Perhaps we were taught that others would try to take advantage of us rather than help us. It seems sure that those who approach life with a poverty mentality end up chasing after the crumbs. Those whose approach to life is more grand seem to receive more than their share of the world's abundance and prosperity.

The really good news is that as cosmetologists we have a truly remarkable opportunity and capacity to contribute in a way that so immediately counts. Our work can revolutionize individuals' appearances and consequently their feelings about themselves and their whole experience of life. What good fortune to have an opportunity to make such a difference for people!

Transformational results don't just happen automatically. It takes real excellence and focus on our part. It's an act of giving. Merely doing a simple haircut doesn't ordinarily make the big difference for people. It's about executing our art with uncommon craft and vision. It's about motivating people to value themselves enough to take time with themselves at home and visit often enough to maintain their look with brilliance.

It's important to hammer home the truth that you are in a profession that gives you the real opportunity to make a profound difference for others. The more people you affect and the greater that impact is and the more often it occurs, the more you set into motion the forces of abundance and prosperity for yourself. The more you give of yourself, the more you receive in return. The more you receive, the greater still is your capacity for giving.

TRANSFORMING CLIENT FEAR INTO HOPE WITH EMPATHY

Don't be surprised to discover that people naturally resist change. They resist expanding their comfort zone, even when it's to their overwhelming benefit to do so. So, just because we're ready, willing, and able doesn't mean that our client is going to let us work our magic free of anxiety or resistance. Many clients will experience fear of change or fear of making a mistake and it will require genuine human relations skills on our part to dissolve their fear and inspire them to proceed. It's in the ability to lead people that we are able to make the greatest difference.

The way to lead people through change is with empathy. Empathy expresses kindly understanding and gentle reassurance. It puts the focus squarely on the client. It takes mental and emotional effort to demonstrate empathy. It's real giving! And sometimes the answer is no. Have the mental toughness and sense of understanding to avoid feelings of disappointment. Remember that virtue is it's own reward. You throw the seed of opportunity without attachment to the specific outcome knowing that the eventual harvest will be abundant indeed.

The more you give, the bigger your dreams. Consider the numbers of people you can effect and influence from a single chair or from an entire salon. With a clear vision of what you can contribute also comes the opportunity to dream big dreams. A well known motivational book bears the title *The Magic of Thinking Big*. A magic comes to you when you give yourself permission to dream big dreams surrounding your capacity to contribute to others.

But here is a reality that is a paradox but that must be driven home. It works in parallel with the law of cause and effect. This is one of the great twists of understanding that highly successful people navigate. And, it's not just a matter of accepting this on a philosophical level. Service to others must come first—and that's in our self-interest. Yet, we must not make the mistake of serving others with self-interest foremost in mind. That won't work. The service to others must be sincere and free of focus on our reward.

What a concept! Can this be true? The information I'm sharing with you is truly life changing. Yet it takes real presence of mind and self-discipline to refine ourselves to operate in this way. Yet, it is the truth behind the universal philosophy of success.

Summary

When we're preoccupied with ourselves, we often focus on what's wrong, what we lack, what we don't have, and how we're incapable. When we refocus outwardly and concentrate on the good we can do for others, something magical happens to our thinking. Here's what we discovered:

- It's healthy for us to define ourselves positively and redouble our efforts toward acceptance.

- Our thinking about ourselves and our circumstances is often a matter of perspective and attitude.

- We can control our thinking and thereby control how we experience life.

- Focusing on the good we can do helps us fulfill our sense of purpose and creates a more positive and healthy state of mind.

Happiness and joyful circumstances and thinking are attracted by our whole approach to life. Accepting life as it unfolds and having a humble sense of what we can contribute leads to the grace of inner peace and serenity amidst the hubbub of civilization's onward march. The teachings of the ages are that we can't go out and get it; we can only prepare ourselves so that it can come to us.

CHAPTER 6

Harness the Great Motivator

> *The pain of progress lasts a season. The pain of regret lasts a lifetime.*

There are two great motivators of human action. The first is pain avoidance and the second is pleasure seeking. We have a fear of pain so we are naturally inclined to avoid pain in all its forms. We also desire pleasure so we are motivated to seek pleasure. That's how complicated motivation is!

But let me ask you a question. Which of these two great motivators is the strongest? Is it fear of pain or desire for gain? It's been demonstrated time and again that avoiding pain is a far stronger motivator than attaining pleasure. A tremendous amount of our behavior is concerned with the fear of pain and the desire to avoid that pain.

WHAT YOU WILL DISCOVER IN THIS CHAPTER

- You'll learn the tremendous power of fear and how it influences your career and life.

- You'll find out about the true nature of the fear of success and discover ways to unhook from negative habit patterns that hold you back.

- You'll see how you can interpret events from a growth-oriented perspective and experience ways to develop a healthier approach to your career in the salon.

- You'll be able to make affirmative choices that enable you to more and better service to your clients.

THE POWER OF PAIN

What is pain? Physical pain is easy enough to define. It hurts! That's the reason we so naturally develop an aversion to activities that can result in physical pain. For example, we naturally develop a scissor technique that avoids cuts and stabs to the fingers and hands. We don't even have to think about keeping our hands off the hot end of a curling iron. The reality of painful consequences creates a healthy fear.

Then there is emotional or psychological pain. This type of pain is often caused by making mistakes, being rejected, not getting what we want, losing what we have, and generally feeling out of control. Most of us have learned not to like these states. As a result it's perfectly natural for us to be afraid of making mistakes. It's natural for us to be afraid of being rejected. It's natural for us to be afraid of loss and not being in control. It's natural for us to be afraid that we're not going to get what we want. It's human to want to avoid emotional pain. So we develop a fear of activities that can lead to emotional pain.

PAIN AND PERSPECTIVE

Some things are painful by definition. If you get kidney stones, you'll experience pain. However, often pain is elective. Somewhere along the line we chose to and agreed to link certain circumstances to being painful. We could just have easily linked these circumstances to growth, progress, and goal attainment. One of the things that separates the successful from the unsuccessful is how they define circumstances. Is it seen as painful or gainful? If the interpretation is one of gain, they automatically dissolve unproductive fears that keep them from growing. As a matter of fact, those fears probably haven't even pierced into consciousness.

It's just like any athlete will tell you, you must allow and experience "growth pains" if you're going to achieve higher levels of athletic prowess. The stress and pain of running harder, lifting more, jumping further must be endured to achieve improvement. Serious athletes constantly stretch themselves physically. They even learn to enjoy the pain because they know it is associated with growth and improvement.

The similar idea comes into play when fear helps us avoid physical danger. It's healthy because it keeps us alive. What virtue is more admired than courage and bravery? Whether it be sky diving or pulling someone from a burning wreck, replacing physical fear with courage and bravery is called valor and it's a highly admired virtue.

Judgment is required to know when to fly in the face of fear. Judgment comes from experience, and experience usually involves making some mistakes. It takes courage to endure the fear and pain so that judgment can be earned. Experience is about learning in the school of hard knocks.

The very same courage is required to endure the emotional pain required for growth. To grow and become more capable as people we have to confront our fears. Fear is the great enslaver. Fear holds more people back than any other force on earth. This is why Franklin Delano Roosevelt's words were so brilliant when he said, "the only thing we have to fear is fear itself."

It's the fear state of mind that is the problem. It's not the circumstances themselves that are so fearsome; it's the emotions that we have attached to the circumstances. Potential reactions and responses to situations are many and varied. They are learned. So, in areas of our life where fear holds dominion, it would do us well to reprogram our responses so that they become productive rather than enslaving.

THE FEAR HABIT

In the hierarchy of response, we move from being inspired, to acting on instinct, to acting impulsively, to being compulsive, and finally to being repulsive. The further down the ladder we go the more fear comes into play. The higher up the ladder we go the more love comes into play. Ultimately, our responses to situations are a choice.

However, many of our responses to recurring situations have become so well worn as to be habits. We respond by habit. Habits become such knee-jerk reactions that we don't bring our judgment or power of choice to bear. Choices are made impulsively or compulsively and without much, if any, conscious thought. Once we form and reinforce a habit it will continue to operate unless it is purposely changed. Quite literally we are held captive by habits that are often not creating for us the results that we want. Yet, almost unconsciously, we continue.

To grow in our capacity, we must grapple with our habits. We must free ourselves, our thinking, and our behavior from compulsive fear-based habits that are holding our careers back. We must go through the pain of change so we can emerge as more capable and fully functioning humans. The effort and energy is worth it because the pain of growth lasts only a short while. If we shirk from it, then we have to contend with the pain of regret, the pain of making excuses for not being, doing, and having all that we could have—and the pain of regret lasts a lifetime.

Let's get specific about some situations when fear of emotional pain is likely keeping us in a rut of unproductive habit patterns. By seeing these situations this way we can begin to see the profound impact that fear is having on our careers and how it is holding us back from success.

FEAR OF "KNOW"

We fear what we don't know. Fear of "know" often comes right down to the fear of making a mistake and a fear of the consequences associated with making the mistake. These fears paralyze us into inaction. Inaction has two consequences. First, there's no growth or development, and spontaneity and "impulse" opportunities are lost. Second, the inaction associated with the fear of "know" means that situations can unravel because they're not being addressed.

Two big areas where these consequences can come into play are technical advancement and business procedures. Technically, many designers are afraid to expand their comfort zones beyond the "tried and true." They end up doing the same old styles over and over again. These cosmetologists have a fear of making a mistake with certain services and consequently avoid them like the plague, even though they could be lucrative and of tremendous benefit to salon guests. Or, if they do the service occasionally, they're purposely limited in the sophistication and creativity they bring to the process. They have a very limited comfort zone and have a fear of expanding it.

I make it a point to visit many salons in my travels. I go through numerous consultations. It has always struck me as remarkable the number of times I've brought up the idea of having my hair highlighted or waved only to be talked out of it! I've come to the conclusion that the reason why this has happened repeatedly is that stylists are not confident about their abilities with these services, are afraid of the possibility of making a mistake, and will refuse the business and decline the money rather than risk the emotional pain of making a mistake. It's a classic example of the fear of "know."

In the next section, we will put the focus on stimulating impulse purchases of unplanned services. So, it's absolutely fundamental that we are free of fear when it comes to performing services that we want salon visitors to purchase on impulse.

Now, let's turn our attention to the other fear of "know" and that has to do with business procedures. Fear of making business mistakes enslaves more salon owners and managers to mediocrity than anything else I can think of. I believe that designers should become fully versed in managing their own careers as business professionals. Remaining oblivious to something as basic as career financial management is done at great peril. The breadth of this fear is so profound that a few broad brush strokes will give you a feel for what I mean here.

1. *Fear associated with numbers and paperwork.* The dislike of paperwork—uncertainty with numbers—lack of comfort with forms—confusion over how to work with statistics and percentages—all of these mean that many salon professionals don't comply with governmental regulations and laws. Books aren't kept or are in such disarray as to be meaningless. There's no financial management of the career or of the business. There's no planning or savings for the future. All of this is disastrous. A life can't be run this way in the modern world much less a career or business that has any hopes of going anywhere.

2. *Fear associated with laws and contracts.* Whether we're talking about building leases or booth leases, people sometimes elect to remain in the dark at their own risk. Perhaps they're afraid that they won't understand or that it's beyond them anyway. Maybe they're afraid of lawyers. We enter into agreements all the time and as citizens and business people we agree to operate legally. Electing to remain ignorant will inevitably lead to a day of reckoning.

3. *Fear associated with "people management."* Whether you're managing the boss or managing the staff, fear can play a fateful role. What a paradox! The staff can be afraid they'll be fired if they make a misstep and the management can be afraid the staff will quit if they do or say the wrong things. This is not a healthy environment. Make no mistake, there can be fear, anger, and resentment among coworkers. These kinds of work-

ing conditions are actually dysfunctional. Everyone is afraid of how everyone else is going to behave and so everyone is walking on pins and needles and measuring their words and there's just an undercurrent of real stress and anxiety.

That's what fear causes—stress and anxiety. In these situations we're talking about being literally paralyzed by fear. It's a state of high-stress inaction. We have fear associated with doing something and fear associated with not doing the very same thing. On the one side of the equation, we have the fear and pain associated with getting the financial house in order and on the other side, we have the fear and worry associated with everything falling apart. And so, when neither option looks good we sometimes elect to do nothing, hemmed in on both sides by fear, worry, and anxiety.

Can you relate to what I'm talking about here? This is not a nice place to be. This is no way to live. It all flows from a fear of not "knowing" what to do. Ignorance may be bliss—but when we know that we should be doing something but are paralyzed by fear from doing it, we are in a state of anxiety. When we get into that loop of "I have to, but I can't—I have to, but I can't," we know we're in trouble and there's no rest until we confront the fear, learn our lessons, and engage the pain of growth.

Let's share a few points to ponder that can provide understanding, encouragement, and direction for conquering "know" fears.

Technically, you can always be confident that you'll do a better job for your clients than they can do for themselves at home. Often that's their other alternative. They have no meaningful training so to them it's a real experiment. They're working with nonprofessional products. If their results are disastrous, they're not at all equipped to do anything about the situation themselves. If we're going to be afraid of anything, we ought to be afraid of the disastrous results clients may experience when they try to cut, color, or perm their hair at home. The same is true for nails and skin care. The general public has neither the knowledge, skill, equipment, or product necessary to do the job on themselves.

As Zig Ziglar has said, "anything worth doing well is worth doing poorly at first." It's the old practice makes perfect principle. Early on in the learning curve of anything we try, we're going to make some mistakes. That's reality. Let me share again that the expert is the person who has made all the possible mistakes. Academic understanding isn't going to get the job done when it comes to an "art" like hairdressing. It's only by hands on experience that we are going through trial and error, to become expert. Resign yourself to this reality. Try to be a fast learner!

The same holds true for business knowledge and processes. We're going to make some mistakes in people management. We may have to learn some lessons in the school of hard knocks when it comes to legal requirements. We may have to burn the midnight oil, confused with the numbers we're calculating. So what! Go to classes. Hire competent professionals to help you and show you. Give yourself permission to do a less than perfect job on the way to improving and ultimately getting quite proficient at these matters.

Accept the worst that can possibly happen and try to improve from there. Dale Carnegie, in his book *How to Stop Worrying and Start Living*, suggested that when we have anxiety about something, to take a deep breath and imagine the worst that could possibly happen. Then accept it. If it happens, you're still among the living so what's the big deal. You lose a customer. You lose a few dollars. You experience a bit of embarrassment. It's not the end of the world. The odds of the worst outcome are actually pretty remote. If that's the worst that can happen, then everything is an improvement from there and that's cause for optimism. Simply going through this process and accepting the worst as a possibility has the effect of dissolving much of the worry. It's worry that causes paralysis.

Give yourself some credit; even if you do a less than perfect job it still won't be that bad. You have education, training, and experience. So what if it's not A+ work! Sometimes you're going to get a B- or a C. Big deal! The earth is not going to stop rotating. If there's a major technical disaster, for example, which, by the law of averages will happen a handful of times in any career, you have your own resources plus the aid of knowledgeable colleagues, to respond. Anyway, little charm and wit and a few nodding heads can a make a world of difference.

Be a lifetime learner. The most successful people never stop learning. They thirst for new knowledge. They understand that their education does not end with their walk down the graduation aisle. The mere fact that you're reading this book tells me that you are dedicated to your professional and personal development. The same principle is true with our experiences. The school of life is to be learned from and we get our most valuable lessons from the mistakes we sometimes make. That "pain" is the price of growth and progress. Give yourself permission to make the mistakes you need to learn from to be able to grow.

FEAR OF "NO"

The other great fear that's so common among salon professionals is the fear of "no" or the fear of rejection. The fear of rejection probably holds down your own income more than anything else. Actually that awareness should be freeing for you because a problem defined is half solved!

However, it's amazing the rationalizations people will make to alter their behavior from the ideal to avoid the potential pain of rejection. A person will know what to say and how to behave in a situation and yet will come up with all kinds of justifications for not doing it. It's fear of rejection. Let's call it what it is.

With respect to your income, let's take a moment and examine the implications of this fear of rejection. It means that you're not stimulating impulse purchasing because your afraid they're going to say "no."

In my seminars I often talk about customer service systems that are consistent and orderly client handling procedures designed to dramatize the excellence of the salon and enhance the experience of guests. Any good customer service system will incorporate a number of opportunities for the client to make unplanned purchases of additional services or home maintenance products.

However, you can have the best system in the world set up and if you have operators who are reluctant or weak when the time comes to stimulate the purchase, you're only going to get marginal results. The fact is, seven of ten purchases made by consumers in North America are made on impulse. It's a sad state of affairs when a salon can get better results out of a cut-out cardboard figure of a human with a cartoon bubble containing a sales message than they can from a living, breathing human being. It's distressing when designers won't say the words because they're so frozen by fear. Or, if they do say them, they're said so ineffectively as to be better left unsaid!

It's not just stimulating impulse purchasing of colors and highlights and waves and relaxers and conditioning treatments. In fast service salons, there are stylists who can't even suggest that they shampoo the client's hair or blow it dry because they're so afraid of rejection. Or how about the fear of rejection associated with these other salon opportunities:

- Designing a home maintenance system

- Demonstrating home maintenance products

- Asking for referrals

- Suggesting gift certificates

- Greeting guests

- Making outbound telephone calls to secure appointments

- Handling inbound phone calls

- Up-selling inbound calls (oh, oh - I used the "S" word!)

- Prebooking the next appointment

- Up-selling the next appointment

- Making a price increase

This is where the rubber meets the road when it comes to making more money. And the only one standing in the way of your making more money is yourself. For your income to grow, you have to grow.

THE "MOMENT OF INSANITY"

In my seminars I often speak of the "moment of insanity." The *moment of insanity* is a psychological twilight zone that we enter into that enslaves us to continue old habit patterns that aren't getting us the results we want. We want different results, like higher income for example, but when the opportunity is there, and the moment of insanity is in play, we're unable to have the presence of mind and fearlessness necessary to proceed accordingly. Instead, with a feeling of uneasiness and regret we repeat the unwanted and ineffective habit patterns. We may even experience guilt and shame at not being able to manage change, which only further drives down our already beleaguered self-esteem.

It's important to comprehend that "insane" habit patterns litter our entire lives. One wise person said that we spend the first part of our lives learning the habits that we then spend the rest of our lives trying to undo. I encourage you to see the scope of what we're talking about here beyond your career.

FEAR OF INITIATING AND CLOSING

Continuing with salon income building as the focus, the moment of insanity rears it's ugly head at two key times. The first time the "twilight zone" can occur is between observing the opportunity to increase income and then proceeding to make the presentation that will accomplish that. Another way of putting that is the fear of initiating a sales presentation. The second time the moment of insanity occurs is after we make the presentation (assuming we've had enough courage to do so) and bringing the transaction to settlement. Another way of putting that is fear of closing the sale. In short, the moment of insanity is the negative head space we can get into surrounding initiating a sale and closing a sale (Fig. 6-1).

In outside sales this is known as "call reluctance." However, in the salon we generally have people calling on us. These salon guests are dissatisfied with their appearance and are looking for

THE "MOMENT OF INSANITY" DYNAMIC

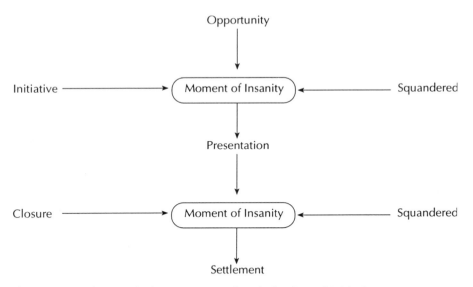

Figure 6-1 The psychology associated with the fear of initiating presenta-tions and closing transactions can lead to countless squandered opportunities and lost sales.

solutions. First-time guests in particular are open to your ideas and recommendations. The opportunity to sell them a smor-gasbord of unplanned services and products is ripe. That's what they're there for! Yet, the fear of initiating sales holds us back from offering our very best to salon guests. The fear of initiating generally boils down to three causes.

1. Low self-esteem and feelings of unworthiness

2. Problems with role acceptance

3. Fear of rejection

We discussed low self-esteem and feelings of unworthiness at some length in the previous chapter. Keep in mind that if we compare ourselves as being less than or subservient to the client we suffer from low self-esteem. I understand that some clients can be demanding and intimidating. Just because a person is older, has an important position, makes a lot of money, or lives in a fancy neighborhood is no reason to become intimidated. They are just folks with their own issues. If they were completely satisfied with how they looked they wouldn't be there to see you to begin with. And, if they start trying to express control over your interaction with them, it's only because they're afraid they're not going to get what they want. It's their fear— see it for what it is.

You don't have to make any excuses for yourself. You're pursuing your purpose in life. We're all members of the same brotherhood of mankind. You're there to make a contribution. You're reflecting virtue and honor as a human being. Stand up straight and tall!

But your feelings about yourself are your feelings. You have to deal with these issues and face your fears head on. A little later in this chapter you'll discover a variety of practical things you can do to overcome your fear of initiating. It's amazing, the self-esteem transformation you'll experience as you move your behavior from being a servant to being a professional.

Soon we're going to delve into understanding the problems of role acceptance and the fear of rejection. Right now, I want you to simply understand that they're part of the background noise associated with the fear of initiating a sales presentation. As you can imagine, you rationalize to talk yourself out of initiating service ideas. For example, we start to think that we have the power to read the client's mind. We begin to convince ourselves that they aren't interested anyway so why bother. We may even take it a step further and think that we'll be appearing pushy or insincere or that they'll think we're just trying to sell them something and take offense. We tell ourselves that if they were interested they'd bring it up in discussion. We go through all sorts of mental gymnastics to justify avoiding the initiation of a sales discussion. So, we initiate no new service ideas and simply take orders as they are made, thank you very much! We simply elect not to act on the opportunity that's in front of us. It's a free choice that dramatically affects your income and quality of life!

Many cosmetologists, however, are able to summon enough courage to make suggestions. But then when it comes time to "close" the transaction they completely fall apart.

See if this scene is familiar. The cutting cape is removed. The client gets up. She was a first-time visitor and everything went so well. You had great rapport. You told her about the various potions and techniques you were using throughout. You took the time to dramatize for her how to get the best styling results at home. As you walked to the front of the salon you came to a little "fork in the road." On the one path you could see the rainbow of home maintenance products glistening and shining. The other fork led to the quick refuge of the front desk and cash register. You knew you had made a commitment to yourself to give clients the full service—to make sure that their visit had the long afterglow a home maintenance system creates. You had planned to design a complete home maintenance system for your new client. You'd done everything to set the stage!

Then suddenly thoughts, truly negative thoughts, started creeping into your consciousness. You started doubting yourself. Your confidence took a nose-dive. The internal dialogue began. "I don't want her to think I'm being pushy. I don't want the good will we've established to be ruined by trying to sell her retail. And our products are so expensive she probably can't afford them anyway. She mentioned she has children so she's probably on a tight budget. I told her about them, if she's interested she'll bring it up. I'm just going to head straight for the cash register and total her service ticket and that'll be fine."

Colleagues, that's the moment of insanity (Fig. 6-2).

I've personally had cosmetologists give me all the ins and outs of the different products that they're using. Then, when they get to the cash register not another word is said. I know that it comes down to fear of rejection.

Alternatively, cosmetologists may present the idea or information with no enthusiasm. They may go ahead and suggest the conditioner, for example, and then do it so half-heartedly and with such a hang-dog attitude that they may well have been better off not even bothering. It's as if they're making the presentation with the expectation of being turned down so they can say

Figure 6-2 Our own psychological problems and worries can interfere with the level of service we offer clients. That's why it's so important for us to govern ourselves by opportunity rather than fear.

to themselves "see, I knew he didn't want it." It's the old principle of the self-fulfilling prophecy. The negative expectation is there during the presentation. The cosmetologist is just going through the paces with no enthusiasm. In these circumstances, the closing question is asked so ineffectively as to practically beg for a negative response.

SYMPTOMS OF THE MOMENT OF INSANITY

The "moment of insanity" has its own symptoms. The most crippling is that we engage in internal dialogue and talk ourselves out of doing what we planned to do during moments of clear thought. We make a decision to do it. Then when faced with the opportunity to do it we talk ourselves out of it. That's insanity!

This little conversation we hold with ourselves is rife with negative anticipation and perhaps anger and resentment. "Why are they making me do this?! I don't feel I should have to do this! They're not being fair with me! I'm not making any money for doing this so why should I bother!" Talk about negative with a capital N!

Furthermore, we have a subconscious file drawer filled with all sorts of negative outcomes from these kinds of interactions in the past. When we face the situation anew, the mental paths back to the negative outcomes are well worn, so in the flash of a second we rush to judgment. If it's something we've had good experience with the judgment is positive. If it has been negative our eyes glaze over with fear of pain.

In the moment of insanity—especially when we're trying to disengage from old unproductive habits to grow more effective—our body chemistry can undergo alteration. Our heart rate, skin temperature, pupil size, breathing, and even our complexion can be instantly effected. Our posture and body language change. Remember, 80% of your communication is nonverbal.

Our voice quality is altered including the rate, volume, and inflection of our speech. Our word choice is influenced—we may even make a "Freudian slip." So, when we open our mouths to communicate, with saucer-like eyes, it doesn't come out very well. We may stumble and stutter or even be at a loss for words.

Customers perceive this lack of confidence on our part. How could they miss it? So, they respond to our energy with some uncertainty. Occasionally, you may have a client who is understanding and realized that you're merely frightened. That kind soul may be very reassuring. However, most consumers are demanding and expect a level of confidence. When they don't see it they have the tendency to decline whatever is being offered. This in turn magnifies our mental state. Our expression worsens. Ultimately, the idea is rejected. We file that away. Who wants to go through all that again?

SOLUTION—ROLE ACCEPTANCE

Many cosmetologists have a problem with accepting the role of salesperson. In our work we play many roles. We're designers. We're hosts. We're artists. We're managers. We're entrepreneurs. We're "psycosmetologists." We're also salespersons. Everyone one of us is a salesperson. You are a salesperson. You may not want to believe it. You may deny it. But it's true. We're going to discuss a number of specific sales techniques later, but we have to accept this role and feel good about it if we expect these techniques to yield the results for us that they can.

I myself am proud to be in sales (Fig. 6-3). When people ask me what I do, I say with pride that I'm a salesperson. Making transactions gives me tremendous satisfaction. I've often heard it said that nothing will develop a human being more than sales. Being involved in sales will make you grow and develop as a person and as a professional. That, in and of itself, is a fantastic reason for embracing this role.

Remember what we learned before—success is not something that you achieve by what you do; rather success is something you attract by who you become. Sales will refine you as a person. It will refine your character. It will force you to confront

Figure 6-3 One of the roles we play in the salon is that of salesperson. We identify problems, prescribe solutions, and motivate actions. That's sales!

your demons. It will improve your communication and leadership skills. Sales will cause you to be more as a person!

By embracing your role as a salesperson, your income will increase instantly. Having worked with tens of thousands of salon professionals over a number of years I can state emphatically that those who cultivate their sales ability and take pride in their sales are the ones who, by a wide margin, earn the most money and enjoy the finest lifestyles and earn the recognition of both their peers and clients.

Some colleagues don't want to accept that sales is one of the roles that they play. Those colleagues don't sell very much. They don't want to accept that selling is part of what they do for a living. They are in denial. They avoid initiating discussions. They only want to take orders. They're missing out on 70% of the purchasing that's being done because of their blind insistence on not growing.

Sometimes I hear cosmetologists justify their denial of being a salesperson by stating that they went to cosmetology school to learn how to cut hair not to become a salesperson. True enough. So too, the doctor went to medical school to learn medicine. The lawyer went to law school to learn the law. Electricians and plumbers went to trade school to learn their craft. But all of them are also salespeople. Whenever you're dealing with the public you're a salesperson. As a matter of fact, many would say that whenever you're dealing with another human being in an advocacy situation, you're in sales!

The moment we move into a situation where we're trying to persuade someone else to take some action, we're selling. Doctors have to persuade people to have an operation or to stop smoking. Plumbers have to encourage people to go along with the wisdom of having their hot water tank replaced, even though it's an unplanned purchase that's not in their family budget.

If we've done a good job of creating confidence and establishing professionalism we can sugarcoat the process and say that we're advising, consulting, or designing. But we're selling! Those are just sweetened ways of saying what we do. But don't let all the talk of "consultation" confuse you about the reality of what's going on here!

AN INDUSTRY IN DENIAL

I'm going to say something very controversial. We have an industry in denial. I regularly see manufacturers, distributors, chain executives, salon owners, and managers knowingly engage in the pretense that cosmetologists aren't salespeople. Realizing that many cosmetologists have a hang-up about their sales function, these industry leaders look the other way, "suspend reality" and go along with the pretense that cosmetologists aren't salespeople. It's almost as if the "S" word (sell, to sell, sales, selling) is a dirty word in the salon and that's it's never to be used when talking to stylists. So these industry leaders sugarcoat it and dance around it and call it everything but what it is.

To me this is an unhealthy situation. It's holding our whole profession back. Until we fully and cheerfully accept our role as salespeople we're going to be constantly deflected from getting at the heart of what it takes to make serious money. Another aspect of the beauty industry is the cosmetic/fragrance/skin care business as it appears in both department stores and "home party" settings. These colleagues have no denial about what they're doing. They have a strong sales ethic. In fact, the sales ethic comes first with product knowledge a distant second.

I love to observe pink "Mary Kay" Cadillacs glide pridefully down the boulevard. Inside is a successful person. Let me tell you something, those people have a much harder go of it than we do in the salon business. Any kind of direct sales like that is a real challenge. Plus, they only have products! Conversely, we have the opportunity to perform needed services that people will need repeatedly and we design and market home maintenance systems at the same time. Plus, we have a real commercial location. We just have so many advantages. But, no matter how great our advantage, as long as we shirk from sales and deny it as a crucial role we must play, we'll be dooming our income to mediocrity.

Those in the profession who have accepted their sales role and have proceeded past the fear of selling, and the inevitable rejection that accompanies it, are the ones making the most money—because they're the ones making the greatest impact on their clients! So, let me state it clearly and unequivocally. Give yourself permission to be a salesperson and embrace that role wholeheartedly. If you're sales skills aren't very good right now, and they probably could use a little improvement, rest assured you will get many of the answers you need to improve as your read on.

THE TRUTH ABOUT REJECTION

The more and more you combat your moments of insanity, the more and more you're going to face rejection. It's a fear that must be faced if we are to grow. The plain fact of the matter is that regardless of what you're offering the SWSWSWN equation will come into play.

Some

Will.

Some

Won't.

So

What?

Next!

No matter what you're offering, the same equation will be true. I don't care if you're standing outside of Bloomingdale's in New York offering $100 bills in exchange for a $1 bill. Most people will not take it. And, their not accepting it says a barrelful about them and their own fears and nothing about you. You could stand on the street corner and offer to exchange $50 bills for $5 bills and most people will pass. You know what I'm saying is true. The person who is receiving the offer is experiencing a moment of insanity. The person is governed by fear rather than faith and is concerned about the pain of making a mistake.

When customers reject good ideas that we initiate, it is either because we did not do a good enough job of building value, providing reassurance, or arousing desire or it's simply because they're afraid of making a mistake themselves. They could be afraid that the service isn't going to turn out right because they've had bad experiences in the past. They could be afraid that they won't like how they look. They could be afraid that others will reject them or that they will lose their popularity. They could be afraid that their "image" and identity will be lost. They could be concerned about damage to their hair, skin, or nails or fearful of allergic reactions. They could be afraid that the change won't be sufficient and that they'll wind up wasting their money.

The consumers experience fear all right! Fears can surface surrounding the purchase of home maintenance systems. They could be afraid that they're being duped in to spending more than what the products are actually worth. They may be afraid that it's the same as what's in the bottle at the drug store. Or they may be afraid that they'll spend their money and won't get the results they want. They may figure that nothing else has done the job so why bother?

It's all fear. It's important for you to understand that the consumer is subject to as much fear as the vendor. You have experienced it as a consumer. I can remember wanting a lap top computer for when I travel. I went into a big computer superstore and was shown the different models. The salesperson suggested the one he figured would meet my needs. Then he asked me how I wanted to pay for it and I promptly experienced a moment of insanity. I was afraid that buying that one might be a mistake. So I said "I'll think about it." "I'll think about it" is just a sugar-coated way of saying no. It's our natural tendency when we don't know enough, that is, when we don't have enough information to say yes!

This is especially true when we're doing something that's new for us, when we're expanding our "comfort zone." There is a natural fear of making a mistake so we just say no. We can't be any worse off by saying no, can we? But, if we make a mistake, we can be worse off by saying yes, true? And, keep in mind what we've learned about human motivation. Avoiding pain is a greater motivator than enjoying pleasure.

Ultimately, our goal in making presentations is neutralizing the fear of the risk of pain, while at the same time magnifying the certainty of gain.

STRATEGIES FOR MOVING THE TRANSACTION THROUGH THE MOMENT OF INSANITY

It's common to have both the vendor and the consumer experiencing a simultaneous moment of insanity. Just as you're experiencing the fear of rejection that accompanies asking them to buy, they can be experiencing the fear of making a mistake associated with making the purchase.

What a situation! But let me ask you this—who is responsible for moving the transaction through the moment of insanity? Naturally, it's the vendor's responsibility. When we're in the role of vendor, we must have the presence of mind, mental discipline, and courage to keep our own thinking and actions straight. At the same time, we have to reassure consumers and soothe any fears that they may be experiencing.

Those who master themselves and master client psychology in this way are the cosmetologists who achieve the most. They make the most money. They have the happiest clients. They're the most beloved. They become the most as human beings because they've learned the art of guiding people to doing what's best.

RECOGNIZE YOUR MOMENTS OF INSANITY

The first thing you have to do is to identify your moments of insanity. You know the symptoms. You're aware of the wide variety of places where fear can come into play. Now, you must personalize this information and take a look at where the fears are affecting you. Become objective to the whole process. Notice the specific thoughts that are emerging in your various salon interactions. Pay attention to when and where moments of insanity are occurring. Recognize them for what they are. The last one you want to fool is yourself so be brutally honest in your reflections. Take a good look.

It's important to get a real feel for exactly what it is that we're dealing with here. Approach it like a scientist with yourself as the subject. Merely identify the mental and psychological phenomenon you're experiencing, but be detached from it. Be an observer of yourself. Don't make any judgments at this point. Just observe what is going on.

This is actually half the battle. For example, observe your thoughts and actions the next time it's appropriate to suggest a $10 conditioning treatment. Do you pull back from the opportunity? What's the internal dialogue that's taking place? What thoughts are racing through your mind? What fears are you experiencing? Are you able to master enough courage to initiate the suggestion or do you talk yourself out of even mentioning it? What client reactions are you anticipating? If you do suggest the idea, what thoughts are running through your head as you talk? What's happening psychologically? How is your vocal quality and word choice? How is your eye contact? What are the thoughts and feelings you experience as you approach the close of the transaction? What's your psychological state? What are your expectations? Do you feel graceful and confident or awkward and uncertain?

See it all for what it is! Experience it. Today or whenever you are in the salon next, watch yourself. Check your thinking and your state constantly. Begin to get a feel for your own personal moment of insanity moments of truth.

ATTACK ONE MOMENT OF INSANITY AT A TIME

The next step is to decide on one recurring situation that you're going to transform for the better. You'll disengage from the old unproductive habit pattern and learn a new way that serves your clients and your pocketbook much better. Now, rest assured that over time you'll be able to approach all your salon fears so you can effect massive change for the positive. But it's best to start with one.

I'd suggest that you start by picking one that really counts. Maybe you're afraid to up-sell chemical services. Maybe you never ask for referrals. Maybe you've got a big hang-up about product retailing. It could be something else, but pick one. Pick the one where the benefits of changing are substantial and where the consequences of not changing are clear.

Another thought would be to pick one where you feel positive about being able to successfully establish a more productive habit pattern. You might as well start with a success. It will get your momentum moving in the right direction.

CALCULATE THE BENEFITS OF SUCCESS

It's good at this point to contemplate the benefits you'll receive by conquering that particular moment of insanity. Let's say, for example, that you're going to start initiating a recommendation of an appropriate haircolor service with each client. In the past you've been afraid, but now you're going to move ahead. What benefits will you enjoy by transforming this habit pattern? Be specific.

Well, you'll make more money. How much more? If you can get one extra color service per day and you work 250 days a year that's an extra 250 color services. If you charge just an average of $40 per service we're looking at an extra income stream of $10,000. If you get half of that then you've increased your income by $5000. And, that's not including tips or repeats! Now what will you do with that money? Be specific.

Another benefit is that haircolor is a loyalty service so you'll retain more clients. Before you know it, there will be more demand for your services than you can supply. You're turning requests away! That's when you increase prices. Let's say in a year you're able to increase prices 15% and still remain 100%

productive. So now instead of making $20,000 you're making $23,000. Instead of making $40,000 you're making $46,000. Get the picture!

There are a lot of other benefits to developing this new habit pattern as well. List them all . Give yourself abundant reasons to change. That way you have a nice "spoonful of sugar to help the medicine go down."

REVIEW THE DISADVANTAGES OF NOT GROWING

On the same piece of paper review all the negative things that won't change if you don't change this habit. The credit card bills stay high. The furniture isn't replaced. You still can't afford to go on that big salon cruise you've been dreaming about. Think about all the circumstances in your personal life that could be relieved with the added income. Contemplate all the personal and professional goals that will be in peril if you don't change. Remember, the more you do of what you've done, the more you'll have of what you've got! Be clear so you know the wages of not changing.

Actually, the dissatisfaction that you have with your life as it is right now provides you with your greatest impetus to change. Magnify that dissatisfaction. Become fully conscious of the long-term implications of staying right where you are. Commit to yourself that all those negative things are not going to happen to you. That you're not going to wind up old and penniless. Your children won't end up without the value of a good education. You won't wind up in an old apartment with rickety sticks of furniture saying to yourself, "if only. . . ."

PLAN A PSYCHOLOGICAL RESPONSE

Because you've observed the emotional and psychological battleground of the moment of insanity, you know that there's a psychological war to be waged. Even though you've made the commitment to rework a particular habit pattern does not mean that the psychological pain is going to automatically dissolve. Be aware in advance that the old habit pattern is going to do everything it can to remain dominant.

Continuing with the haircolor example, as you move into initiating discussions, the old thoughts are going to rear their ugly heads. In fact, they may even become more intense. They may magnify to attack the very premise of change. These thoughts will try to talk you into remaining the same and not growing. Stay aware of this fact. When you move into the heat of battle the negative thinking will come into play. Forewarned is forearmed, right?

You want to plan a psychological response. Proper preparation will promote a positive performance. Distill and create positive thoughts before you enter the heat of battle. Make them affirmations that reflect the positive value and virtue of your new actions. Make your thoughts first person and present tense. I'd recommend that you write your affirmations down and practice them in advance.

> I am a professional. I care about the success of my clients. I am committed to motivating my clients to look their very best. I understand that when my clients feel good about how they look, their whole quality of life improves. I have an opportunity to improve my clients' psychological outlook. Haircolor does improve appearance. Haircolor makes this client look and feel better. I am a committed professional. I have the self-discipline and presence of mind to proceed as a professional.

When you enter the "twilight zone" of insanity and the negative thoughts start flying, have the awareness to see the reality of what's happening, the presence of mind to shift into affirmative action, and the mental toughness to stop the negative thoughts and insert the positive affirmations you had planned. That's how you take psychological control of the situation. When you review the affirmations in your mind, many negative thoughts may try to wedge their way back in. Don't let them. It may be painful for you to repeat your affirmative thoughts—you may not even fully believe what you're saying to yourself—but do it anyway! As time goes on the negative thoughts will completely dwindle away and the positive thoughts will become second nature.

Another good psychological strategy is to visualize yourself successfully handling the interactions in advance. Literally see your consultation in advance and observe yourself thinking the positive thoughts and fending off any old negative thinking successfully. Just see yourself proceeding with composure, grace, confidence, presence of mind, and experience an overall feeling of happiness and joy.

PLAN A PRACTICAL RESPONSE

As we're engaging battle with our insane behavior patterns, mental activity and psychological tension will be present. It's a good idea to have a practical tool capable of releasing the anxiety from the situation. If we're just standing there "naked," with only our words to rely on, we may wind up being more nervous than necessary.

However, if we bring some preplanned device into play, it can become the focus of attention and consequently relieve much of the stress and anxiety. You want to specifically know what you're going to do and say in advance. Plan it out. Practice it. Don't be at a loss for words. You'll be able to refine it and improve it over time but have the professionalism and discipline to create preplanned presentations. Don't just "wing it." In the next section you'll get many ideas for planning what you're going to say. Right now the key is understanding that if you have a preplanned presentation you can distract yourself from a lot of the insane thinking because you have to focus on the presentation.

Devices like stylebooks, your own portfolio with before and after pictures, brochures and salon menus, service description sheets, consultation forms and prescription pads, haircolor swatches, foils, caps, and color brushes are examples of the kinds of devices you can use to focus attention. By incorporating these tools into your presentations not only will you give your consultations more sparkle, but you'll have something to hold onto and talk about, which will release the tense energy and help keep insane thoughts at bay.

MAKE A SMALL PROMISE TO YOURSELF

The moment of insanity that you've decided to attack—start your attack right away with your very next guest. The only thing you have any power to influence is the present moment. And change is now. Don't put it off. If you put if off for a day you may put it off for a year. Your future is going to be nothing more than the culmination of all your present moments. Make a small promise to do the very best you can at engaging the decided-on moment of insanity with your very next guest. The only thing you have any power to influence is the present moment. Yesterday is history. Tomorrow is a mystery. All we have is the gift of to-day, which is why it's called the present!

Just make a little promise to yourself and keep it. It's that simple. It's just the next client. You can do it the one time. Even if you stumble all over your words and make mistakes, don't worry. Don't worry so much about how you'll do. You'll do fine. Even if it's not polished and graceful that's OK. The key is that by acting immediately, you're in the solution. By remaining stagnant, you stay in the problem. The key is getting started. Don't just think about it—do it! Don't plan to do it—do it!

Adopt the "one client at a time, one day at a time" philosophy. Keep the transition in small bite-sized chunks. All you have to do is promise yourself that you'll do it with the very next client—no matter what. Then do it! Even if it turns out that you're next client is a 60-year-old gentleman banker—do it. If it's a regular who has scoffed at haircolor in the past—do it. If it's a 14-year-old boy—do it. Don't look for excuses not to do it. Do it one client at a time.

Be vigilant. Have discipline in your approach. You know that discipline is an act of self-love. You may never have realized that before. It's about saying to yourself that "I'm taking my career and my life seriously. I count. I matter. I'm going to hold myself to a high standard on this." If you start making even little compromises you'd be amazed at how everything can start to slowly unravel. Be hypervigilant.

If you do revert to the old pattern and succumb to the moment of insanity, dust yourself off and get back on track immediately. Don't let it fester. Don't wallow in it; don't beat yourself up. Make a new resolution at once. Do not allow the old pattern to get its claws back into you. Don't backslide more than a step at most—and immediately charge ahead.

ACKNOWLEDGE YOUR PROGRESS

Just the fact that you've started to confront a fear that's holding you back is a significant move. At first, how your salon guests respond is of secondary importance. Naturally you'll want to polish and refine your approach. After each interaction ask yourself what you could have said or what you could have done that would have made the interaction more productive. Make yourself a constant, nonstop self-improvement project.

But the surprise of surprises will be when you discover that some of your guests will take you up on your suggestions. Remember the SWSWSWN equation. No matter what, you can be sure that some will. Likewise, we've discussed that some won't. So what! It bears repeating that we must not concern ourselves so much with the some who won't. Our interest is in unearthing the some who will. They're the ones improving your quality of life. They're the ones giving you the affirmation. They're the ones telling you that your opinion counts, that you matter—that they're interested in what you have to suggest. The real tragedy is when we become so afraid of the rejection of the some who won't that we don't give ourselves the opportunity for the acknowledgment and affirmation of the some who will!

Our goal is to polish our presentations to the point that a larger and larger percentage of people will and a smaller and smaller percentage of people won't. Because of the tremendous number of people who visit a salon in a week or month we can always look to the "next" one to improve on. We have a steady stream of people to practice on so it's realistic for us to make measurable progress in reasonable time. What more can you ask?

Take note of your progress. Take pride in your advancement. It can be a source of great satisfaction. It's a real self-esteem builder when you can actually watch yourself growing and developing as a person and as a professional.

Make the decision now. Use the transformation of that one habit—that one moment of insanity—as the first battleground where you're going to wage your war for a better life! Make it real important to yourself. And, step up to it. Don't be lazy—don't put it off. Use it as an opportunity to take control of your life and move your life and career in a new direction.

Summary

Now here's what it comes down to. You either accept the pain of change, which will only last a short while, or else resign yourself to accepting the pain of regret that will last forever! Here's what we discovered:

- Your greatest psychological motivator is the fear of pain, and pain is often just a matter of perspective.

- Fear of pain runs rampant in the salon and often holds us back from offering the very best we have to salon guests.

- By conquering our fear of selling and rejection, we allow ourselves to become more fully functioning and giving professionals.

- Presence of mind and a plan of action are indispensable tools for making progress as professionals.

Change involves pain—emotional and psychological pain. But the pain only lasts for a short while. With growth comes freedom and the opportunity for new growth. You become a more magnificent person. Not changing, allowing fear to continue to govern your activity, leads only to regret—the regret of lost opportunities—the regret of wasted time—the regret of having to make excuses to explain failure when there was plenty of opportunity for success.

A day at a time, a guest at a time—your future is in your hands.

PART III

ACQUIRING YOUR FULL MEASURE OF SALON WEALTH:

Implement Success Systems to Attract and Retain Salon Clients While Maximizing Client Spending Power

CHAPTER 7

Why Aren't All Salon Professionals Rich?

> **To earn like a professional you have to behave like a professional.**

We can look up to many who've achieved remarkable success in our industry. However, we can also point to too many who are not doing so well. Even though all the positive conditions are present, many people in our business are just scraping along. This is true in any career, but it's so unnecessary in our profession.

Every year tens of thousands of people leave cosmetology because they don't know how to make enough money to survive. I've seen many talented individuals leave the profession over the money issue. On the other hand, I've seen colleagues with only average ability do remarkably well because they've mastered the art of making money. And it's not just designers. Many salon owners are barely able to keep their doors open. Every year thousands of salons go bankrupt. Thousands more are put on the market because their owners aren't finding it profitable (and

161

may, in fact, be losing money). And don't forget that this is in an environment of extremely high consumer demand.

WHAT YOU'LL DISCOVER IN THIS CHAPTER

- You'll learn that making money and technical excellence are two separate issues.

- You'll find out about practical things you can do to enhance your professional image while making your consultations more influential.

- You'll come to grips with the reality of time management and teamwork in the salon.

- You'll be able to avoid some of the most common mistakes that doom salon careers prematurely.

MAKING MONEY IS ITS OWN THING

Let's clear up one matter right away. The ability to skillfully perform services and the ability to make money are two separate issues. In fact, your ability to perform beauty services is only going to contribute about 20% to your success in the cosmetology field.

This entire section is devoted to the specifics of making money. Your willingness and ability to master the skills and methods discussed is going to make the difference between poverty and prosperity.

FAILING TO CAPITALIZE ON DEMAND

We're in an environment of booming demand for our products and services. What are the major reasons cosmetologists fail to capitalize on this? Let's discuss the big three:

1. *The inability to attract clients.* You can have all the skill in the world but if you have no one to perform it on you're dead in the water. To attract clients means that we must be promoters and marketers. We must have strategies in place all the time to keep a steady stream of first-time guests visiting us. Experience teaches us that it's best for us to focus on low cost/no cost strategies, which means that we will have to put in some serious effort. If we just put out the open sign and wait in the salon hoping that people will just walk in, we're not coming to grips with reality and we'll wind up in the poor house.

2. *The inability to retain clients.* It takes time, effort, and money to get customers in the door. My own research tells me that it costs about $25, $50, or sometimes more to get a first-time visitor in the salon. Generally, the customer has to visit three or four times for the salon to break even on its marketing investment. If we can't keep clients coming back, we're in trouble.

Why don't people come back? Naturally, if the quality of the work isn't there they won't be returning. But beyond that, indifferent service is the top reason people don't return. On the other hand, remarkable service creates the impression of excellence, quality, and value. That will keep them returning. Don't underestimate the importance of friendliness and personality and the demonstration of personal interest. Those characteristics are the hallmark of successful salon professionals everywhere.

Often stylists and salons don't have a strategy to keep people coming back. Client retention, as we'll discover, amounts to more than merely giving your business card at the conclusion of the service.

3. *The inability to maximize client spending in the salon.* How the beauty professional consults with the client is the biggest variable in the amount of money the client will spend. Take the very same client visiting for just a haircut. One stylist will ring up $20 at the end of the visit. Another stylist will ring up $70. What makes the difference is what is said to the guest.

Two stylists can be at the same salon on the same day for the same number of hours and service the same number of patrons, yet yield completely different financial results. The only difference is that one actively stimulates impulse purchases of additional services and products and the other does not. It's fantastic because increasing client spending only requires a few well-placed words and actions.

Be willing to be proactive in this process and you'll end up with a lot more to show for your time spent in the salon. Maximizing client spending requires effective and focused consultation. The stylists who openly share ideas and information with guests and enthusiastically encourage clients to avail themselves of what's being suggested are the ones who make the most money. It's the difference between being an order maker and an order taker.

SALON PROFESSIONALISM

In the next chapters we will detail ways to attract and retain clients as well as ways to maximize client spending. Before we do so, however, we must delve into the basic issues of professionalism. Professionalism is at the very core of achieving success in your salon career. A number of practices will, on their own strength, either put us in position to enjoy meaningful income or completely take us out of the running.

Salon guests must like us and trust us if we're to be successful. The more professional we are, the more clients we will attract and the more they will spend with us. Our consultations will be more influential. Our reputations will be more esteemed. We have to show we are serious about our profession and there are several things we can do enhance our professional image. Some of these items seem very basic; however, they must be reviewed because they are fundamental to achieving success.

BASIC PROFESSIONAL HABITS

Show Up on Time Ready to Go. It seems basic, but it's certainly worth emphasizing, that we must show up for work every day. The people who make the most money in the salon are the people who are there day in and day out. Do you know how many times the championship performers call in absent every year? Zero times! They're there every day! If they don't feel 100%, they still show up. If the weather outside isn't ideal, they still show up. If business looks like it might be slow, they still show up. They've developed an attitude of responsibility and reliability.

You experience a tremendous amount of personal pride when you know you're one of those people who can be relied on to perform as promised. It's amazing how something as simple as dependably showing up every day attracts the respect of our colleagues and clients. It's easy to understand that something as basic as this is always a prerequisite for personal and professional advancement.

Those who look for excuses to avoid reporting for work often have a problem making money. If they feel a little sick, they call in sick. Feeling a little sick is not an excuse for failing to report to work. We all get aches and pains, suffer headaches, and feel the blahs from time to time. Always have enough self-respect to grin and bear it and enough professionalism to have the attitude that "the show must go on."

Sometimes the weather isn't good, or the car breaks down, or we have an argument with our spouse. Some people will use these kinds of situations as excuses for not going to work. If you haven't realized it already, you'll soon come to grips with the fact that the people who grapple for excuses to fail will always find them. However, in the final analysis, while they're disrupting the smooth functioning of the salon, the ones they're really hurting most is themselves.

I recommend to owners that if one of the operators starts to develop a problem with absenteeism they should immediately begin looking for a replacement. The salon and clients deserve the basic courtesy of dependable people who are going to be there to perform every day. And if people can't be relied on to do something as basic as show up, then they're not very serious about succeeding. It's been said that 80% of success in life is simply showing up.

Tardiness Is an Income Killer. We must also make a commitment to ourselves to show up on time. And on time means that we're ready to perform on time. That means we actually arrive 20 to 30 minutes before we're scheduled to start. So if the salon is scheduled to open up at 9:00 o'clock, we arrive 20 minutes early, already put together, to make sure that the coffee is made, the garbage emptied, the towels folded, the lights and music are on, and we're ready to go when it's show time!

There's an extra measure of self-confidence you experience when you're punctual. You relieve yourself of a tremendous amount of damaging stress and anxiety when you're an early bird. People who are late inevitably experience anxiety and compromise their self-esteem. Not only does your own psychological system get beat up, but you run the risk of becoming cross with colleagues and coworkers. When latecomers play catch up during the day it's easy for them to lose focus and begin rushing through clients in a frantic attempt to make up lost time. None of this contributes to making more money because it stands in the way of exercising relaxed and focused attention. It all can be avoided! That's the rub!

I can remember once visiting a salon on a consultation project. The governor's wife was on the premises making telephone calls and rearranging her schedule because her stylist was late. True, the weather outside was snowy. But the client was there on time. The salon owner was there on time. The rest of the stylists were there on time. I was there on time. The stylist finally showed up more than an hour and a half late claiming poor driving conditions. Needless to say the professional reputation of the salon was irreparably harmed by the misdeed of one.

Consider how the poor stylist must have felt. Situations like this often have as their root cause a low self-concept. If the designer had a vision of herself as a "can-do" championship performer, she would have simply not accepted anything less than arriving on time no matter what the weather. If it meant starting the commute early—so be it! But if she did not have those feelings about herself or, conversely, had feelings of resentment, anger, or justification ("they're lucky I'm even showing up!"), then the situation has the sad result of reinforcing the low self-esteem and the cycle continues to spiral ever downward.

If your time management hasn't been all it can be, then at least make this resolution right now. Do yourself and everyone else a favor and promise yourself that you'll forever more avoid the basic mistake of stumbling in bleary eyed and incoherent at 1 minute to 9:00 AM—still needing to style your hair and apply your make-up. Treat yourself and those around you with more consideration than the mini-crisis which 5, 10, or 15 minutes of tardiness can create. You know how you feel when clients are kept waiting and the salon manager is steaming—don't put yourself through it. You deserve better than that for yourself!

The "When's My First Appointment" Phone Trap. Of course, successful stylists don't fall into the trap of calling in late either. It's a fast path to low income. I've had situations where the stylist calls in at 5 minutes to 9:00 and asks, "When is my first client scheduled?" When they hear the first client isn't going to be there until 11:30 AM, they announce they'll make their appearance at about 11:00 AM. This type of spontaneous schedule adjustment simply cannot be tolerated! It's not professional, it lets down the whole team, and if it's a pattern in a salon, it's a real indication of a business spiraling out of control.

Whether it be an appointment-based salon or a walk-in establishment, the champions are always there the full slate of hours to catch any extra business that might happen along. When we work at a salon it's a basic commitment we make to be present during the agreed hours. It's really quite simple. The thought that "we're only being paid when we are actually doing customers so why should we be forced to be there just to sit around" is truly small-minded. First, many general activities are required just to be prepared for guests when they do arrive. Second, being a good team member means that we're there to support our colleagues even when we're not busy ourselves. The idea that we're too good to help others or won't help colleagues because we're not being directly paid for that function is actually quite selfish. It's not the hallmark of any championship performers I know. More on teamwork later.

A Strong Response is Appropriate to Combat Tardiness. Ultimately, salon managers must set the standard themselves and must nip tardiness in the bud and not allow it to continue. There has to be a strong policy to discourage lateness. The consequences must be meaningful to regulate those who don't seem to have the maturity to discipline themselves.

Managers need to respond to these situations instantly. The moment the offending operator walks into the salon, the issue must be immediately addressed. If it were to happen in my salon, this is what I would say in private to the offending party:

> Your success and earnings are very important for me. We want to be able to give you as many fresh customers as possible so your earnings are good. So, if you're only here when you've got appointments booked, that can't happen. The reason why we've agreed to hours that you're expected to be here is so we operate like a team. I've found that I get all my errands done Tuesday morning when I'm not scheduled to be at the salon. You'll find you can get all your errands done during your times off. Also, if designers started calling up and changing their hours suddenly, you can understand that it would be a mess. My goal is to make sure we serve our clients well and that all the designers are able to earn a decent living. So, it's really important you make sure you're here during your scheduled times. Does that sound fair? The last thing we want is to have to issue a formal warning that could jeopardize your next promotion. Thanks for letting me share.

The first principle of professionalism is showing up on time every day ready to perform. You'll always feel good about yourself when you know you're consistently reliable.

Always Look Your Best at the Salon. You'll find it interesting to know that people judge you within seconds of seeing you. It's true! They judge how much money you have, your upbringing, your level of education and sophistication—even how honest, moral and ethical you are. The list goes on. These are primordial instincts that have remained with us since caveman times. We size people up—friend or foe—fight or flight!

Now the good news is that we're in the image business. Our whole training has to do with helping people create an image that makes them feel good and look good. However, it's important to come to grips with the social implications of fashion and image and realize that people generally perceive that folks of different social classes present themselves differently. The ladies at the country club appear one way, and the waitress at the local diner looks another. You get the idea.

And, it's not only how one looks. It's also how one speaks and acts and walks and eats and behaves. Now this is a powerful insight to come to grips with, especially when you consider that money has a way of chasing those who already have it (or at least appear to have it). So my suggestion is that you, who are already an image and design expert, use your knowledge on yourself and package yourself for financial success!

The salon professionals who make the most money always look great. This covers everything from the clothes they wear to their make-up to their grooming to their hair. In most salons, operators can make their own fashion selections. The better you dress and the more effectively you present yourself, the more money you'll make. Studies have shown a direct link between how you present yourself and your income. I'd suggest you present yourself as if you already had a lot of money (Fig. 7-1).

Clothes Make a Big Statement. You'll find it to be of great financial benefit to strive to be a fashion role model for your clients.

Figure 7-1   As an image professional, you have the knowledge and ability to package yourself for success.

Take a look at the customers visiting the salon and dress a little bit better than they do. And not only must the clothes be fashionable, they should also be well maintained. And don't forget the shoes. Beautiful footwear makes a strong positive impression and demonstrates your attention to your appearance. Make sure they're polished and shined. We're in the fashion business and we're offering fashion services. Let's work to create a little excitement about fashion by how we present ourselves.

Gentlemen should wear neatly pressed dress slacks and a buttoned down shirt. I know many male designers who wear a jacket and tie and it makes great sense. Ladies would do well to wear a dress or skirt and blouse outfit. In some salons, the operators wear a uniform. That uniform must look spectacular. It must be clean and pressed. It must appear fresh.

Sometimes new stylists will make the excuse that they can't financially afford to dress well. I don't believe that for a minute. If it takes going to used clothing stores and selecting slightly worn fashionable garments then so be it. We simply have to dress the part for our income to be that of a professional. It's not a case where we make more money so that we can dress better—it's that we dress better so that we can make more money!

Pants worn by female operators are not recommended. I've discovered in my own salons that if an inch is given on the issue of wearing pants, a mile will be taken. So, I simply recommend not wearing pants. I understand that it may sound like a bit of a double standard, but I'm simply reporting fashion and cultural reality. Also, jeans of any color, sweatshirts, T-shirts, sneakers and the like are also not professional salon wear. They may be comfortable, but you'll end up paying dearly for that measure of comfort. This type of clothing is simply too casual for a fashion environment. It doesn't matter how expensive the jeans are or how fancy the sweatshirt is, it's simply not professional attire and it will affect on your income. Sexy and revealing clothing is also inappropriate.

Meticulous Personal Grooming Is Essential. Make-up, if you're going to wear it, should be fashionable, well applied, and understated. If it means taking lessons from someone who knows the art of applying make-up, then do it. In fact, many salons are now offering make-up services and lessons. There's something stylish and glamorous about properly worn make-up. It's one of the delicate areas of fashion that distinguishes ladies who know how to finish themselves properly.

Sometimes it's a good idea to ask other people for their honest feedback. What you must avoid is the painted face look. Painted faces look cheap and diminish the professional image. Heavy make-up can cause a person to look hard and low class. This is something to be avoided at all costs in the salon. Often we get into the habit of certain make-up early in life and neglect to update ourselves. Keep in mind that people judge us instantly by our appearance and we need to package ourselves effectively. Fragrance and jewelry are also worthy of attention. Wear make-up and fragrance in the manner of people of substance. It makes a great impression.

Our personal grooming requires meticulous attention. Naturally, we shower and wash every day. Use deodorant. Gentlemen must shave daily. We endeavor to keep our skin clear. It's important to keep our hands and fingernails as clean as possible. Teeth must be brushed regularly and breath kept fresh at all times.

Keep in mind that we're in close proximity to other people and must not risk causing any offense in our personal grooming. Be extra careful on this matter because people will notice, but they often won't speak up. Hold yourself to the highest standard of freshness and cleanliness all the time.

Be a Hair Model Every Day! Naturally, our hair must be meticulous and fashionable. To guarantee this, it must be completely styled and set in place before arriving for work. I can remember a young female stylist arriving for work just in the nick of time with her hair washed but unstyled. It's not surprising that walk-in customers were already in the reception area waiting to be served and she simply didn't have 10 or 15 minutes to get herself pulled together. We were so busy that day that she ended up working her entire shift looking miserable, and I know she was completely preoccupied with her stringy unstyled hair. Clearly, this is not a very professional image for a stylist to present to salon clients. I've seen this scenario repeated time and again, and it's not the standard of success you want.

Dress up to go to work and create a favorable first impression and you will automatically glow with the shine of success. Remember, we're in the fashion business and how we appear can motivate others to look their best. All this will affect your pay immediately.

Use Time Management and Teamwork. To achieve the highest level of income one must be a good manager of time, That not only means showing up at work on time, it also means working a full day and staying on time throughout the day.

Common time management mistakes among salon workers include taking long breaks away from the salon, taking long lunch breaks, leaving work early, and refusing customers at the end of the day. These are all serious errors that ultimately have an impact on income.

Leaving the Salon Early is a Big Mistake. It's common sense that we remain at the salon throughout our scheduled hours of work and that the salon be open for customers when it's scheduled to be open. You'd expect that of any type of business. Think of times when you've rushed to the bank or post office to get there before closing and how frustrated you'd be if they decided to lock their doors 10 or 15 minutes early. Or, consider how disappointed you'd be if you arrived at a restaurant at 10:30 PM after an evening movie for a quick bite only to be informed that the kitchen was closed and the cook left early because it was a slow evening. You expected them to be serving until 11:00 PM!

The public expects a business to be open and ready for service during its posted hours. Is it too much to ask?! Businesses that fail to deliver on this basic level not only miss out on new customers, but lose existing customers as well. Don't think for a minute that salons are immune to this reality of the marketplace. No matter what, we must be open and available during our scheduled hours.

Imagine this scene. It's Saturday afternoon at 2:30. Jennifer just finished her last scheduled client. She notices that there are no walk-in clients waiting in the reception area so she figures this would be a good time to leave a little early. After a quick look at the appointment book she approaches the manager and declares "I'm all done with my clients for the day. I have some errands I need to run so I'd like to leave a little early if that's OK. Barbara doesn't have any one else scheduled so she can handle any walk-ins although it looks slow right now. If that's OK, I'll see you on Monday."

This type of request creates a difficult situation and shows a lack of consideration on the part of the operator.

1. The fact that the designer is without client is exactly the reason she should linger. It's apparent that they could use a few extra customers and access to a potential walk-in or two can only help in this cause.

2. It puts the manager in an awkward position. Let me assure you that the manager would like to deny the request. Unfortunately, if the operator is denied premature leave of their work duties it's quite likely that they will be long faced and brooding the rest of the day, which is simply bad for the environment of the whole salon.

3. Let's consider those walk-in clients. Have you ever noticed how often they arrive in twos or threes or fours? And, how they can arrive quite suddenly? And, how they all want to be serviced right away? It doesn't take much imagination to recognize that if one of the designers leaves early, the business could be compromised and the speed of service jeopardized. In fact, the whole deal could be lost.

Professional businesses remain open during their scheduled hours and professional operators are available to serve guests during the hours they promise to be there.

Keep to Your Schedule. We must do our level best to stay on time with clients throughout the day. The old days of keeping people with an appointment waiting are long gone. Doctors, lawyers—all the professions have improved on this count. So too have salons. Clients are very time sensitive and simply will not be kept waiting. If the guest is scheduled for 2:00 o'clock in the afternoon, we make sure we're ready to serve them at 2:00 o'clock. It's as simple as that! Proper booking, working with efficient speed, and using teamwork all help to keep us on time.

Offering "impulse purchase" services to clients is one of the best ways to increase service revenues. This may tighten our schedule a little bit. Fine, so we work rapidly. The stylists with the highest earnings are inevitably efficient workers in a team environment. So we always do our best not to fall behind and that means we work quickly. That does not mean that we sacrifice the quality of the service or our attention to our clients. It merely means that we pick up the pace and stay focused on the task at hand.

If we are unavoidably running late with no opportunity to catch up, then simple courtesy requires a telephone call to upcoming clients to inform them so they're not kept waiting endlessly when they arrive.

Teamwork Makes the Difference for Higher Income. We work as a team in the salon. TEAM means Together Everybody Achieves More. The most successful salons apply a team approach to customer service. The team approach is real simple. You only have to keep in mind two principles.

1. Be alert and notice when something needs to be done.

2. Offer to help before being asked.

To be a good team member means that we have to be paying attention to what is happening in the salon. When we have a couple of free moments, the first thing we do is ask ourselves,

"What can I do to assist one of my colleagues?" We scan the salon, notice if anyone is running a bit behind schedule, and then we proactively offer to help. It's that simple.

For example, we see stylist Brenda just getting started on the blow dry for Mrs. Brown. We notice that Brenda's next guest, Mrs. Adams, is nervously paging trough a styling book in the reception area. We immediately conclude that it would really help Brenda get caught up if we shampooed Mrs. Adams hair and prepared her for her service. So, on our own, without being prodded, we quietly approach Brenda and ask if we can help and get Mrs. Adams ready for her. We ask if there's a special shampoo or conditioner she would like us to use. Once given the go ahead, we approach Mrs. Adams, introduce ourselves, and explain that we're going to give her a luxurious shampoo and get her ready for her appointment with Brenda.

That's what teamwork is. It's about seeing when and how we can help our colleagues and then proactively offering to help and giving it our very best. It's all a part of helping each other with time management and providing the best service to the customer. Salons that know and practice teamwork are always the most successful. It's a culture of helpfulness that's contagious and is appreciated by coworkers and management but most of all by clients.

Teamwork Is a Two-Way Street. Naturally, when you proactively help colleagues you create an environment in which they will help you when you're in a pinch and they're available to help. It's mutual support. But not only that, it keeps clients in the salon.

Let's say that later Brenda goes on maternity leave. Who do you think Mrs. Adams is going to request in her absence? She's going to ask for that stylist who gave her that fantastic shampoo a few months before. Or, let's say Brenda is overbooked on a particular day and Mrs. Adams calls and simply must get her hair styled. Who is Brenda going to recommend? Brenda is going to suggest a colleague who Mrs. Adams is familiar with. It only makes sense.

Let's look at the other side of the coin. What happens without teamwork. Operators become overly possessive of customers.

It's almost like it's a capital offense to even exchange a few words with a salon guest who's not your client. This is not positive energy. Time management is sacrificed. When operators get behind, they stay behind to the detriment of their clients. The positive opportunity to encourage "impulse service purchases" is diminished, negatively affecting everybody's income and hurting service to clients. Client service is also sacrificed because operators are so possessive they don't even think of booking one of their clients with a colleague in a scheduling pinch. The ability of a salon to retain clients in a stylist's absence is diminished because the atmosphere that allows for this is nonexistent. Salons without a culture of teamwork often breed selfishness, jealousy, anger, resentment, and even cross words among coworkers. It's not an environment where the client comes first. There's often too much negative energy. The lack of a team ethic is simply not conducive to achieving high levels of income and prosperity.

As a matter of fact, to promote the reality of teamwork, I'd recommend that a sign communicating the following message be put in clear view in the reception area or washroom: "WE ALL AGREE THAT A CHANGE OF STYLE OR STYLIST CAN BE REFRESHING. PLEASE FEEL FREE TO REQUEST ANY HAIR DESIGNER AT ANY TIME BECAUSE WE WORK TOGETHER AS A TEAM TO PLEASE YOU."

If a few clients want to hop from designer to designer from time to time that's OK. It's far better for them to stay at your salon rather than feel they have to go someplace else to try something different. Everyone will make out fine in the end, but most importantly, more clients will be retained in the salon.

Avoid Client Turn-Offs. We want to impress our salon guests and not present ourselves in a way that may be disagreeable. Here are some basics to keep in mind.

Salon Cleanliness. First and foremost is salon cleanliness. The salon and your station must be meticulous at all times. The next time you're out of town visit a top salon and observe their level of cleanliness. I think you'll be impressed. It's easy to understand that in an environment where people are experiencing personal grooming services that there be a high expectation of cleanliness. To the public, anything less than spotless is not acceptable.

Have you ever been seated at a table at a restaurant that has not been cleaned? Quite a turn-off, wouldn't you say? I can't tell you how many times I've seen a salon guest escorted to a station that is an absolute pigpen—hair all over the place, wet towels chucked against the mirror, bottles and cans and scissors and tools in disarray. This disorganization and sloppiness does not create a feeling of confidence!

The reception area, the back bar, the rest rooms—all public areas of the salon have to be kept neat and tidy at all times. We clean and straighten as we go. We don't wait until the end of the day. This all seems basic; however, the standard of cleanliness is compromised in salons too often.

The fact is, that it's everybody's job. And if everybody starts thinking "anybody but me" can do it or that sooner or later somebody will come along and do it—then what winds up happening is that nobody does it. Sometimes you'll end up doing more than your share. But don't let that stop you.

Many salons develop a schedule of who's responsible for what. And that's fine as far as it goes. But inevitably some tasks are going to fall between the cracks. And an attitude of "it's not my job" solves nothing. You want to be a part of the solution and don't forget that your guests are being serviced in that environment too! The champions simply take the lead and do the work that's in front of them without a feeling of anger or resentment but rather with a feeling of quiet joy in the knowledge that what distinguishes them is their ability and willingness to do what others won't or can't do. Virtue is its own reward!

Personal Behaviors That Can Detract from Our Professionalism. There are five basic areas of conduct worth looking into because they have the potential to make a strong negative statement to salon guests and diminish our professional image.

1. *Smoking.* In the days of Humphrey Bogart and Lauren Bacall smoking was fashionable. No more. In fact public attitudes toward smoking have turned noticeably sour. Smoking is now

regarded as low class by the mass of the population. That's just the truth. It certainly does not have any connection with fashion or glamour or beauty, I'm sure we can all agree. Consequently, we don't smoke in front of clients. We don't smoke outside the front door of the salon. We don't smoke anywhere that a client could see. And, if we do take occasional smoke breaks, we wash our hands and freshen our breath before we return to serving clients. People can smell us if we don't, and many people will be deeply put off by the lingering stink of tobacco. We certainly don't want to risk offense.

2. *Chewing gum.* Sometimes when people don't smoke, they chew gum instead. As a matter of fact, some of the gum manufacturers advertise gum chewing as a substitute for smoking. Well, once again, gum chewing does not carry an image of glamour or style. Unfortunately, it's part of the negative stereotype some have of designers. That's particularly sad because chewing gum is not considered sophisticated, fashionable, or beautiful. Chewing gum in front of customers will only bring up negative, counterproductive images. As a result, we simply don't chew gum in front of clients.

3. *Food and beverage.* Simply put, it's a bad idea to eat or drink in full sight of salon guests. The front desk, the station, the reception area, the back bar, and other public areas of the salon are not appropriate places for eating our lunch or taking a coffee break. It just doesn't seem sanitary and it gives the public the wrong impression. You wouldn't find it happening in another professional business and we need to hold ourselves to a high standard too. I wouldn't even recommend having a can of soda in view of clients. Most salons are equipped with at least a small private staff area and that's where our nourishment in the salon takes place.

4. *Drugs and alcohol.* One of the biggest problems facing business is alcoholism and drug abuse. I know it's a problem because I've had to deal with these situations first hand. Suffice it to say that drug and alcohol use have no role whatsoever in a professional salon environment. I know that some salons occasionally celebrate with a little champagne or cordial for the guests and that can be fine as far as it goes. What we're speak-

ing specifically about here is the use of drugs and alcohol by staff and management.

Let's be straightforward. We don't imbibe liquor at the salon or during or before working hours. We don't smoke or sniff or take illegal drugs at the salon or before or during working hours. If you or a colleague feel the need to drink or use drugs before or during working hours then a very serious health problem is present and help should be sought at once. This is not a matter of humor nor is it something to be taken lightly. If drug or alcohol use has spilled over into a person's work life then the individual could well have crossed the line to addiction.

Drug and alcohol addiction gets progressively worse over time and addicts will ultimately sacrifice everything (including a career) to the object of their addiction unless they receive help. Help is available. To start with, any number of 12-step programs and treatment centers offer immediate assistance.

A word to salon managers. Use good judgment whenever socializing with colleagues outside of the salon. Use these as opportunities to build the esprit de corps. You set the standard of what's acceptable. I know of situations where the manager has gotten involved in heavy drinking and drugging with her staff and it's a very bad scene. If you "party" with your staff like that you're setting the stage for a drug and alcohol culture to seep into your business. Believe me, that's not what you want. You must demonstrate discipline and composure on these outings.

5. *Awkward communication.* People judge us by the words we use. Although it's nice to have an extensive vocabulary, at the very least we want to make sure we don't damage our standing with poor language or usage.

- Grammar. If a person uses poor grammar it's like they have a sign on their forehead that says "stupid" or "uneducated." It's really quite sad because poor grammar will hold a person back in life. If you have any suspicion whatsoever that your grammar skills need improvement, then don't wait another day before acting. A simple way to start is to ask a couple of well spoken friends to correct your mistakes, and then work hard to make the changes they point out. You may want to consider taking a class or getting some tutoring. You must make sure you speak with proper grammar or everything else you do to make a positive first impression will be sacrificed.

- Foul Language. We never use foul language, even if the client uses it. You know the words I'm talking about. And, if you have any doubt about a word, simply don't use it. The fact is that we immediately downgrade ourselves in the eyes of others as soon as we utter a foul word. Foul language is low class and no matter how well we've presented ourselves, as soon as a nasty word is spoken, it's like a foul blemish. And once the word is said, its impact can never be undone.

- Dirty Jokes and Off-Color Stories. These are simply not professional so we don't tell them. Likewise, we don't make racial, religious, or ethnic comments. We don't put anybody down or make fun of any group in our society. It's simply in bad taste and it doesn't put us in a very good light.

- Spiritual Discussions. The salon is not the place to hold spiritual discussions. I've had managers come to me about designers who are so full of religious zeal that they talk about their faith with practically everyone who sits in their chair. People come to a salon for beauty services and it's just not right for us to question people or preach to them about religious matters. I know that intentions are well meaning, but the salon is not the proper venue. Such discussions can potentially turn some people off to the extent that they'll never visit again.

  Also, it's not good form for designers to discuss intimate details of their personal lives with clients. Medical problems, relationship issues, and other matters that aren't the business of the client and have nothing to do with the reason for their visit are better left unsaid. I don't even think it's a good idea to put up pictures of the family at the station or display other items of a personal nature. It's just not highly professional.

- Gossip. One thing to definitely avoid is gossip about other clients or stylists. Gossip almost always has a negative edge to it and amounts to little more than character assassination. Maybe 30 or 40 years ago the salon was a place where the dirty laundry of the community was aired. But that day is long gone and must never return. If you gossip about that guest who just left, imagine how the customer who is sitting in your chair must worry about what you're going to say about her when she leaves. You can't successfully have that kind of energy floating around.

- Criticism. Another basic no-no is criticizing other salons. Have you ever heard doctors or bankers criticize fellow professionals? They wouldn't think of it. Our objective in conversation is not to point out the trials and tribulations of colleagues, but rather to share the positive and exciting things we're doing professionally. To repeat a point made earlier, we refrain from insulting the haircuts our first-time guests received at their former salons. I've seen stylists comb the guest's hair down over her eyes and make disparaging comments. It's humiliating and insulting for the client. That kind of a ploy is one of the most unprofessional and thoughtless maneuvers we can pull.

- Advice. Finally, the notion that part of the designer's job is to offer psychological advice to salon guests is nonsense. Naturally, we're very free with appearance advice, and make loads of suggestions on how clients can look and feel better about their appearance. But we leave it at that. Keep the relationship strictly professional. The best strategy is to keep the conversation focused on hair. Discuss the products and techniques we're using and share useful information that can help the guest at home.

Stay at One Salon Long Term. If you want to achieve and maintain a high level of income the best advice is to stay at a good salon for years and years. It's true that the people who make the most money and achieve the highest status in our industry are the ones who have longevity at a salon. Contemplate for a moment how serious professionals in any field establish a track record of long service at one location. The doctor I went to as a child is still practicing from the same office. So is the dentist. So is the optometrist. The winners make a career of it and hang in long term.

One of the reasons this practice is of such benefit is that it encourages supreme confidence from the buying public. People are creatures of habit and place a high value on dependability. This is the reason referrals flow so naturally to veterans and why the scope of clientele naturally progresses from one generation to the next.

Another reason long-term employment is of such value is that the veterans are always there to pick up the wayward customers left behind by operators who come and go. When you start combining these factors it's easy to understand that once designers have been at a shop for more than 5 years, they're to the point where they are making some very handsome returns for the time they're working—often in the range of $20 an hour or more—and that's in the 1990s.

On the other hand, migrant workers suffer low wages! The most basic mistake that cosmetologists make is moving around from salon to salon. Almost always there is an immediate loss of income. As a matter of fact, I'm not aware of there ever being anything but an immediate loss of income where independent salons are involved. It usually takes months and months to recover financially from the cost of a transition. Even if the new salon is offering a more generous pay structure, the bottom-line income loss remains.

Also, moving around establishes an unprofessional and costly habit pattern (Fig. 7-2). Every salon has its challenges. The trick is to be a team player and work through problems and find positive solutions short of departure. Developing a habit of running away from difficulties, rather than meeting them head on, sets people up for following this pattern throughout their professional and personal life. Any worthwhile goal will have obstacles to be overcome. Running away is not the answer.

Sometimes a cosmetologist has a better opportunity in another salon. That's fine as far as it goes. But how about the idea of mak-

Figure 7-2 Moving from salon to salon creates career instability and usually results in a downward income spiral. The best strategy is to stay at one salon long term.

ing a better opportunity for yourself where you are currently? You're a player on your current team. And you're the most important variable for your own success. It may be better to be a contributing member of a team reaching for the heights rather than a second stringer in an already established salon. Think about it.

Making Threats to Leave Is a Poor Strategy. Sometimes operators will make a series of threats to quit in an effort to maneuver issues to their favor. Their concerns can be anything from where their station is located, to whether or not they receive a promotion or pay increase, to the hours they're required to work. In discussions, these operators make the veiled threat that if things aren't settled to their liking, perhaps they'll pursue their career elsewhere.

This is a counterproductive strategy. It shows no faith or loyalty and demonstrates a complete lack of compromise and team spirit. The final result will often be the exact opposite of what the ones making the threats want. They may win the battle, but they'll lose the war. An astute manager will sometimes give in to keep the peace but then immediately begins looking for a replacement. Another likely result will be for the manager to de-emphasize income growth opportunities for the operator making threats. For example, a walk-in requests a spiral perm. The manager is likely to give the opportunity to anyone but the one making threats of imminent departure. It only makes sense, don't you think?

Do Your Best to Retain Your Job. Previously, we mentioned a whole list of issues that relate to basic professionalism. Violating those tenets repeatedly will lead to job loss. But there are a couple of other items that weren't mentioned that need to be addressed here.

First, don't talk down the salon or its management. A negative attitude will lead to dismissal. And, the better the salon you're in, the more quickly it will lead to dismissal. There is a high positive premium for team members who support and encourage the management and even show a little gratitude from time to time. Those are the people who get promoted and will receive preferred assignments. On the other hand, those who make it a habit of bitterly complaining and criticizing everything and everyone are usually shown the door in short order. Nobody wants to be around negativity, and an astute manager will remove a bad apple before others are contaminated.

Second, respect the property of the salon. Certainly, it's clear that we don't steal anything from our place of employment. We don't walk off with perm rods, nail files, color tubes, bottles of product, or anything else. We don't "borrow" things from the salon or our colleagues without clear permission. If you think that lifting a perm from the dispensary to do a friend at home is a fringe benefit of the job—think again. It's theft and it's a crime. Salon managers keep a pretty accurate inventory and know when things are missing. When there's shrinkage, inevitably a spirit of distrust arises, which penalizes everybody, including clients. Ultimately, anyone who's involved with pilfering is eventually discovered and dismissed.

The bottom line is that you have a great career ahead of you so just be sure not to make any foolish, basic mistakes that can sidetrack you.

Be Careful About Pay Cuts that Are Disguised as Pay Increases. Sometimes a few percentage points on the commission are enough to persuade an operator to make a move. This is usually very short-sighted thinking. On the surface it seems like a good idea. Deeper investigation reveals otherwise. Consider that the profit margin on salon services is only somewhere in the range of 7%. Hardly a fortune. And an increase in pay rates has to come at the expense of something else. Maybe the quality of the products used for services is affected. Maybe the advertising doesn't get done. Maybe there's no money for a receptionist or salon assistant. Maybe customer amenities are sacrificed. Maybe ongoing education is short changed. All the things necessary to enhance a bigger business and ensure client satisfaction are trimmed back. What inevitably happens is that the business shrinks and the income of all suffers.

Take a moment and really come to grips with what's being said here. The higher the percentage of revenues paid to operators, the less that's available to keep the business competitive on other fronts. Consumers are smart and know when they can get more elsewhere so the client base is gradually eroded and overall revenues slide. As revenues go down, pay inevitably declines. So, ultimately, the higher the percentage of revenues that goes to payroll, the lower the revenues become and the less the bottom line pay amounts to. What would you rather have, 40% of $1000 or 60% of $500? The answer is that 40% of $1000 ($400) puts you with more money in your pocket at the end of the day even though the percentage is smaller. What you're aiming for is higher income not a higher percentage. Lower percentage often results in a higher income because more funds can be devoted to building and retaining customers.

So, as you can see, the higher commission brings with it a price that usually results in an ultimate loss of income and opportunity. It's an amazing paradox that many of the highest producers with the steadiest incomes work with what many may perceive is a relatively low rate of commission. Investing in client traffic, building promotion and quality service extras ends up yielding a bigger paycheck at the end of the day.

The way to increase income is to serve more clients and increase client purchases. That's the formula that has always worked. Trying to increase income by increasing commissions simply is not the solution.

Don't Blame the Salon When It's Up to You to Create Income. Sometimes stylists want to make a transition because they've not been able to make enough money where they're currently working. Why? Are these stylists taking enough responsibility in the clientele creation process? Or, are they just sitting around waiting for walk-ins? Do they have a problem retaining clients? Is the cosmetologist offering high quality service and personal attention?

Avoid Making Sudden and Emotional Transitions. Sometimes colleagues may encourage you to make a move with them while they're making a transition. This is a maneuver to be skeptical about. Odds are they really aren't thinking about you, they're thinking about doing maximum damage to the salon they're leaving or scoring points at their new salon. So be very wary!

If your salon has been purchased, don't let that be a sign to move. Stay and work with the new owner and manager and try your very best to make a go of it. If you show a little bit of flexibility, understanding, and cooperation, your loyalty will be highly prized!

The bottom line is that job transitions send you back to the starting line and cut your income at once. The key is to select the right position and then have enough character and determination to make it through life's challenging moments. Keep in mind that anything worth accomplishing is going to have obstacles. Running away from problems to "start fresh" can be a negative habit pattern that will hold a designer back from high levels of income.

Summary

The marketplace pays rich dividends for professionalism. If we want to receive a professional level of compensation, we have to hold ourselves to a rigorous professional standard. Here's what we discovered:

- Making money is its own thing and requires a set of skills beyond technical excellence.

- Showing up on time ready to go, looking the part, using effective time management and teamwork, and avoiding common client turn-offs are fundamental to success.

- Staying at one salon long term is the most basic way to experience income growth and earn a reputation for dependability.

If you simply commit yourself to these basic principles, you'll avoid the most common career mistakes. You'll be laying a firm foundation for achieving happiness, success, and prosperity in your career.

CHAPTER 8

Attracting Salon Clients

66

They can't buy if you haven't got them in the door.

99

You must keep a steady flow of fresh clients coming to you. People move on, move away, and pass away. You must replace them with new clients. Also, it's simply wise to be continuously building demand for your services. That way the law of supply and demand will create a situation allowing you to augment your income by price increases. Of course, if you're relatively new to the profession or new to an area you must build an abundance of client demand to enjoy economic prosperity. All of this means that you have to keep a fresh flow of customers coming your way.

WHAT YOU WILL DISCOVER IN THIS CHAPTER

- You'll learn how to make the most of your personal contacts and put to use the power of word of mouth to create an endless flow of referred clients.

- You'll find out about salon phone techniques you can use to convert more first time callers into regular clients.

- You'll discover how to use incentives more effectively to produce extra salon business.

- You'll feel more confident and be more effective interacting with walk-in guests.

- You'll appreciate how you can network and cross-promote with colleagues and other merchants to generate more salon commerce.

CULTIVATE YOUR FAMILY, FRIENDS, AND ASSOCIATES

The easiest and most economical way to get fresh clients is to cultivate your family and friends. Let me dispel the idea right away that you perform services on family and friends for free. I just don't believe in that. Maybe you do your mother for free but that's about where it ends. Your extended family and friends ought to be your first line of clients. And you charge them a fair price. It's as simple as that. You don't need to make it any more complicated or go on any kind of a guilt trip over it.

If your friends and family expect you to do it for free, you'll simply have to explain that the matter is out of your hands. Say that it's the salon policy and you have no ability to change it. That way you don't have to take the heat yourself.

If you're the owner, and family and friends are expecting free services, you'll also have to explain that it's the salon policy that everyone pays. Even though you have the power to make exceptions, it would be a very poor example to set for the designers working there. Say that you learned long ago that this was a rule that couldn't be violated without risking many unwanted repercussions.

You should try to cultivate your family and friends as paying clients. That's your easiest clientele to attract.

ENCOURAGE PEOPLE YOU DO BUSINESS WITH TO VISIT

The next group to invite are the people you trade with. The people at the laundromat, the lady at the florist, the clerk at the drug store, the TV repairman—whenever you make a purchase of any kind make it a point to mention what you do for a living and invite them to visit. To the lady at the dry cleaners you could say, "I'm a professional hair designer and I notice your hair every time I come in. Your hair must experience a lot of stress in the midst of all the chemicals and humidity that you work with. I'd just love to have you visit so I could give your hair a treatment that would prove very refreshing and relieving. Let me give you my card."

This is always a game of numbers. When you talk to ten people you'll get two or three to visit. The trick is to talk to as many people as you can and let the law of averages do the rest of the work for you.

This is also great practice for approaching people and introducing yourself as a professional. The public is interested in hair design. Be proud of what you do. Have confidence in the value you can bring to others' lives. Step right up and introduce yourself and share your desire to serve.

Make Sure You Have a Business Card. A word about your business card before we proceed. You must make absolutely sure that it contains all of the following basic information: your name; the complete salon address including city, state, and zip code; and your telephone number including area code. Additionally you could provide a complete listing of the services you provide. Your hours of operation are always helpful and you can add "additional times are available by appointment" so that if your hours are limited people won't be discouraged from calling. If there's a distinction or membership you'd like to announce, it can be mentioned on your business card.

CLIENT REFERRALS

Your next line of activity for client attraction is to encourage client referrals. You want to invite people who are currently visiting you to send in their family, friends, and work and social associates. Client referrals should provide a constant stream of new guests throughout your career so it's worthwhile to become an expert at generating client referrals.

UNDERSTANDING THE DYNAMICS OF CLIENT REFERRAL

First of all, convince yourself of the value of these clients. More important than anything else is the fact that they've been presold on you as a professional. Whoever is sending them has obviously had some nice things to say about you or they wouldn't be there in the first place.

My gut feeling is that a referred client is easier to get involved in purchasing impulse services and home maintenance products. These are "make-over" type customers. Also, I believe that they're easier to retain than the average visitor if for no other reason than the fact that it's easier for you to build rapport because you have a common acquaintance. Finally, because it was referral that brought them in, they may be more inclined to keep the referral chain going (Fig. 8-1).

Some clients will be good referral sources and may be able to send you five or ten or more fresh guests. Other clients won't send you any even though they're as loyal as can be. Ours is not to wonder why this is, but simply to accept this reality. Furthermore, we never really know which ones are going to be the great referral sources. You just never know! So, you have to

Figure 8-1 Stimulate "word of mouth" by consistently asking for referrals and you'll cultivate a steady flow of clients.

approach everybody who visits with the proposal that they send you clients. Let the law of averages take care of it from there.

MAKING REFERRALS HAPPEN

No doubt you've had clients recommend their friends to you. However, chances are that's happened purely by accident. With a little bit of consistent effort on your part it can become a daily reality. The key is to regularly cultivate referrals from your guests. You'll want to have a preplanned approach that you execute consistently and flawlessly with each visitor. Also, you'll want to be flexible enough to identify opportunities to ask specifically for a referral. Finally, you'll want to keep the ball rolling by staying focused on regularly reminding clients about the possibility of referral.

REFERRAL PROGRAMS

Many salons provide clients with an incentive to participate in a referral marketing program. That's great. If you're going to do that sort of thing, I'd suggest you be generous.

Keep in mind that it can easily cost $25 or more using traditional advertising to get an average fresh customer in the door. Contrast that with the certain confidence of referred customers and it's easy to understand the extra value. Simply the fact that they have received a favorable report by a trusted friend is of tremendous importance. Also, there's a level of familiarity because you have a mutual associate.

Consider the annual value of a new customer. It can easily be in the hundreds of dollars, with a lifetime value potentially in the thousands. Any referral incentive you offer, no matter how generous it appears, really amounts to no more than dropping a nickel to pick up a dollar!

You might want to give an incentive right away with the very first referral to show immediate reward and encourage additional referral activity. Anything from a complementary hair revitalizing treatment to a favorite bottle of shampoo, to a complete cut and style, to a straightforward dollar value gift voucher can be appropriate incentive.

USE A CONSISTENT "FIRST-VISIT" APPROACH

Approach guests on their first visit about referring people to you. Don't wait until they've been there four or five times before you bring the subject up. The fact is that they may never get there that many times. Keep in mind that the most dramatic change to the guests' appearance generally occurs on their very first visit. That's when their family, friends and coworkers are going to notice most and make the most compliments. That's when the situation is most prime to cultivate referrals. Make sense?

Additionally, the fact that you get them involved in sending you referrals actually helps retain them as clients. While they're talking about you to others they're at the same time reselling themselves on you. As people they know start to drift in, they have a vested interest in your success. They want everything to go well and for you to succeed. This is a great environment for client retention.

When and how you ask for referrals is crucial. During the visit, while the service is going on, I'd suggest making a couple of comments about how others are going to react to the new look. "Wow, that color is really hot. Your date is going to go wild about this." "This cut is really bringing out a very intelligent and powerful look in your eyes. They're going to think you got a promotion back at the office." You get the idea. Start to paint the picture that others are going to notice what's being done here.

Another good stage-setting strategy is to casually weave in an example or two about people who have been referred to you and how great everything turned out for them. You can come out and say that "I'm getting so busy now that I'm only able to take on new clients who are referred to me by my existing clients." Or, "Practically all the new guests who walk through the door were sent by one of my existing clients. It's something I'm very grateful for." Basically you want to create the idea that it's a common and acceptable thing for people to send you their friends and associates and that it works out happily for everyone.

All of this is to prepare the client a little in advance of your direct approach. The best time to start the direct approach is just at the moment of concluding the styling but before the cape is removed. As you show them the finished look in the mirror ask them "Do you like how this looks?" Most of the time you're going to get a strong affirmative or a very slight correction. Then, as you're putting your hand mirror down you look them straight in the eye and say "May I ask you to do a favor for me?" They'll either respond with an immediate yes or else ask you what the favor is. With either response you say the following:

> You're going to get comments on how nice your hair looks, you wear it so well. Most of the new people who visit me for the first time do so because they've seen my work on someone who just naturally shows it off so well. And I don't mind telling you that it's always a special joy for me to serve a friend of an existing client. Let me ask a favor of you. Whenever someone mentions how nice your hair looks, would you please let them know where you had it done and even suggest that they come on in for a visit themselves? I'd sure appreciate it.

It's important to note that you have their complete and undivided attention as you say this. Expect that everyone will say yes because they pretty much all will. Then say "well while you're getting yourself changed I've got something I want to give you so I'll meet you up at the front desk."

When they get to the front desk start with a question asking them if they'd like to receive the main incentive you're offering as part of your referral program. For example, "How would you like to receive a haircut and style on your next visit compliments of the house?" Expect an affirmative answer, then continue,

> Well, it's yours as our way of saying thank you for your participation in our referral rewards program. It's easy. Whenever someone mentions your hair simply let them know where you had it done and give them one of these special menus. It even has a new client offer to encourage them to visit. We'll keep track of who visits. Here's a list of the rewards you receive. The nice thing is that even if only one person comes in you still get a complimentary treatment and scalp massage and

the rewards build from there. It's a good idea to keep these menus handy in your briefcase or purse or car so you have them when you need them. Does that sound fair enough?

Notice how I included an incentive for the new visitor. It's always a good idea to reward them for giving you a try so that they understand first hand how the dynamic works when you get them involved in the referral chain.

You'll want to develop some sort of a tracking system to monitor this. It can be as simple as a "gift certificate" stapled to your menu, which the referred guests bring in that you file under the name of the referring client.

Keep in mind that consistency is the key. You simply make this an ongoing system for new client creation. Understand that some clients will do nothing with it. Some clients will refer one or two. And some will send you many. You never really know which ones are going to do what so you don't prejudge. You give everybody the opportunity to participate with the faith that some of the seeds you plant will yield a healthy harvest.

LOOK FOR REFERRAL OPENINGS ALL THE TIME

You'll be surprised how many opportunities to create a referral are coming your way right now that you're not paying attention to. Whenever guests mention someone they know who could possibly benefit from your service you want to immediately ask for the referral.

For example, if a client mentions a friend or coworker who's experiencing any problems with hair, nails, or skin, you immediately ask for the referral. That's one of the reasons why it's a good idea to keep the conversation focused on hair, nails, and skin and your services because then these opportunities will come up more frequently. Pay attention and look for these openings.

Let's say you're applying hair color and mentioning a technique you're using that helps prevent fading. In the midst of this the client mentions a coworker who is always having problems with hair color. You then interject, "Well, have her come in and see me. Because a lot of professional women come to me, I've perfected some techniques that help avoid that problem. What's her name?" See how easy it is!

Keep Promoting the Referral Idea. Make your big play for referrals on the first visit like we described. But don't just drop it there. Some people have to hear about it several times before it finally sinks in or before they finally relate it to their own personal circumstances. That's the power of repetition. Also, you have a whole army of existing clients who probably have never been strongly approached to send referrals. You'll want to cultivate that resource for sure.

A good place to consistently remind clients of the referral opportunity is in your thank you note. It's a very positive practice to begin sending thank you notes to each first-time guest. There's practically nothing that will uplift your image of professionalism in the minds of clients more than a thank you note sent immediately after their first visit. It also helps retain clients wonderfully.

It's simple; all you have to do is pick up a couple of packages of thank you cards and some stamps and you're ready to go. In the card you could write: "I really enjoyed having you as a guest and styling your hair. It turned out great! I'll bet you're getting lots of comments. Thanks in advance for your help in letting your friends know about me. You can be sure I'll do my best when they visit."

And send your thank you cards out every day. Promptness counts. You want your guests thinking referral while their visit is still fresh. It's a mistake to wait for a week or more and send out a whole bunch at once. Do it the day of the visit.

Making a follow-up telephone call is a great alternative to the thank you note. It's appropriate after any chemical service. Again, that communication presents an opportunity to cultivate referrals. "I was thinking about how great that color looked when you were here the day before yesterday. I'll bet you're getting a lot of comments on it, you wear it so well! I want you to know how much I appreciate your letting everybody know where you had it done."

Of course, the ultimate time to ask for more referrals is immediately on the heels of whenever you're sent a referral. Let your referring client participate in your success. That way they'll take more of an interest in helping by sending you even more referrals.

The best strategy is to thank your clients by phone the day the referred guest visits. You could send another thank you note, which is fine. But a call is just a little more immediate and personal. In the conversation you could say: "Mrs. Smith, your friend Mrs. Jones was in today. What a delightful lady! Thank you for sending her to me. She was all smiles when she left. I really appreciate your helping me in this way and I want you to know that you can be confident that all your friends will get the VIP treatment when they visit. And, you know you're only one referral away from a complimentary service on your next visit. I'd sure like to show my appreciation to you in that way. Thanks again for all your help."

That's all you have to do. Simply drop the suggestion that more referrals are welcome and that you've not forgotten your promise. When people are treated with this level of attention they are pleased to send you friends. And, you'll be able to thoroughly cultivate referral sources. In fact, you'll start a powerful chain reaction!

COLLEAGUE REFERRALS

Depending on the type of salon you work in, you may have colleagues who can refer clients to you. Specifically, those who are providing noncompeting services or who have an area of specialty other than yours. It's really important to develop a team approach for mutual benefit so that clients can be referred within the salon. That way they can have the convenience of enjoying all their needed beauty services and treatments under one roof.

THE "TAG TEAM" STRATEGY

It's a good idea to start by exchanging services with salon colleagues to help set the stage for powerful recommendations and
immediate introductions. Let's say you're a nail technician
working in a salon that also has hairstyling. Trade services with
the your colleagues with an eye toward each being a fantastic
model for the other. Have one cut your hair, one perm your hair,
and another color your hair. And make sure your hair looks fantastic all the time.

Likewise, do a different style of nail on each stylist you're
working with to show off your very best work. And make sure your
colleagues' nails look fantastic all the time.

Then, exchange signs (which are professional) that are
prominently displayed at everyone's station. For the colorist you
could have a sign at your table that says: "Notice the beautiful
lustre I'm wearing in my hair. Ask me about it!" And then at the
stylist's station you can have a sign that says: "Notice the hands
on this working girl. Ask me how I keep them looking so great."

It's best if your sign has some intrigue to build curiosity.
That way your signs will prompt salon guests to ask questions.
When they do you have a fantastic opportunity to make heart-
felt recommendations back and forth among colleagues.

MAKING AN EFFECTIVE INTRODUCTION

Someone comments on your hair color. You say: "Doesn't Janet
do a great job? You know she seems to take every advanced haircoloring class that's available. She really takes the time to design a beautiful look and her results are always excellent."

Be sure to coach each other on how to make the strongest recommendations so that everybody knows how to respond effectively when clients inquire. A colleague could say something like:
"Doesn't Nancy do a fantastic job!? She's so on top of doing nails
and attends all the classes. She's a real specialist and I'd even use
the word 'art' to describe her work. And she has a knack of recommending and designing exactly the right kind of nails that fit
perfectly with the lifestyles of all her clients. She's really super!"

When you've finished your service for the client who has inquired, make it a point of actually walking her over to your colleague and arranging for an immediate consultation. You could say: "Here, come with me for a second. Let me introduce you to Janet who designed my style. Janet, let me introduce you to Mrs. Smith. She commented on how much she liked my haircolor job and I told her how great you are. Mrs. Smith, why don't you sit down for a second and let Janet wave her magic wand and share some ideas with you. I'll leave you two together."

What a great example of how to do it properly. In coaching your coworkers on making introductions, you'll probably want them to follow a very consistent procedure. You'll want them to bring the guest right over to your area. You'll want them to introduce you both and mention the clients interest. They should pay you a compliment and have the client sit down so you can spend some time together alone.

To be most effective you'll want to strike while the iron is hot. So if at all possible hold a consultation on the spot at once. Then, ask for the opportunity to proceed immediately. To conclude your consultation you could say something like: "Your new nails will take only X amount of time, and the total for the whole service is only $Y, that's all. As a matter of fact, the way we have the schedule arranged right now there's no reason why we can't do this immediately. Does that sound fair enough?"

Some will say yes and then you just proceed. Some will say that today isn't convenient for them so you ask, "when would be a good time for you?" You pull out your appointment calendar and schedule a time in the immediate future that works.

Some won't be interested or will want to think about it. That's fine. Don't push and don't feel bad about it. At least you've planted the seed and because they'll be making future visits to the stylist they'll remember you're there when they have a change of heart. You might even give them a bounce-back incentive to keep them warm until their next visit.

If you're really busy, at least take a moment to establish rapport and share a quick idea or two. Then mention you'd like to spend more time with her for a thorough consultation to learn more about her so you can design something perfect for her.

Then, appointment calendar in hand, you suggest alternatives until you arrive at a near-term time (perhaps later that day) that works for everyone.

We call it a consultation because it's a more sugarcoated way of making an appointment. The guest need not have any anxiety about a consultation. The experienced among us know, however, that a consultation is just as good as an appointment and should practically always result in an on-the-spot service. Avoid the mistake of merely giving your business card and suggesting they call if they're interested. That will not get you the results you want.

These colleague referrals need to be planned and executed flawlessly and consistently. With a little coordination you and your colleagues could easily refer a dozen or more quality prospects to each other each week. That has tremendous value if for no other reason than it's easier to upgrade existing clients to additional services than go out to the marketplace and pull in altogether new people. It's also a lot less expensive. When it's both more effective and less expensive you have to favor this kind of an approach. This one technique alone should bring you new clients that will spend thousands of dollars with you each year.

TRANSFORM TELEPHONE INQUIRIES INTO CLIENTS

Practically every salon I know of gets dozens of calls each month from potential clients inquiring about services. The vast majority of these calls are mishandled and consequently a tremendous value slips through the fingers. Seven or eight of ten telephone calls ought to be converted to appointments. My observation tells me, however, that as many as eight or nine of ten calls are lost by the average salon. I'm talking now about calls that you're already receiving. Improving your skill at handling the most common phone-in situations will enable you to cultivate more clientele.

People are phoning in for one reason and one reason only. They're in the market for our services. They want to buy what we have to offer. The main reason we lose so many opportunities is that we're not fully conscious of the opening that we're being presented with and how to take full advantage of it.

It's important for us to learn to regard the call itself as a buying signal. Also, the questions that the callers ask are buying signals. Probably the most common question we're asked is "How much does it cost?" Unfortunately, many of us have begun to link this question with rejection. That's unfortunate because when people start out with that question it's most often because they have no earthly idea what else to ask.

RESPOND TO THE PRICE QUESTION WITH CONFIDENCE

From now forward keep this rule—never disclose price without first building value. The truth is that people need to be reminded that the real measure of value is what they'll receive for the dollar spent. So, just throwing out a blind number without first relating what is received in exchange amounts to no less than giving incomplete information. When people don't have enough information to say yes they will inevitably say no.

If you've been in the habit of disclosing price without building value, then the majority of the conversations have ended unsuccessfully. It's because we've allowed ourselves to be placed in the role of "information booth." Our appropriate role is to be an appointment setter and to use techniques that will make appointment setting a reality.

When you're on the telephone, the first time you hear the price question I want you to ignore it. That's right, pretend that it wasn't even asked. As an appointment setter there is information you want to get, questions you want answered, and a first impression you want to make! Also, the fact of the matter is that the price of your services is of secondary concern to the caller. Whether or not they'll be comfortable, how they'll be treated, whether or not you're friendly—those issues are their real concerns. Simply blurting out some price doesn't get the job done. You must engage them in some conversation to get their defenses down and build rapport.

TURNAROUND IS FAIR PLAY

Understand that the one asking the questions and giving the directions is the one in control. So take control of the situation. That in itself will alter your mental state and improve your performance. Plan in advance how you want the conversation to unfold. Here are three questions you can use to immediately respond to "How much does it cost?"

1. "Is it for you?" That instantly changes the nature of the discussion because it puts the focus on the prospect. That's where the focus should be anyhow.

2. "Have you ever been a guest at our salon before?" That's a legitimate question. Generally the answer will be no and it perfectly sets up what you'll ask next.

3. "It's delightful that you're interested in visiting us. May I ask, what prompted you to call our salon in particular?" That's a powerful question because it's going to generally yield an answer that will boost your confidence as well as dramatically improve your "inquiry-to-appointment" ratio.

You'll be surprised and reassured to discover that these calls are not just appearing out of the blue. Believe me, folks aren't opening up the local yellow pages and calling up all the salons in town to do a price survey! Rather, they've heard about your salon, or have seen your salon and have positive feelings about visiting or maybe they're looking at some advertising material that has appealed to them. Often they've been referred and they just want to make contact before they decide if they want to give you a try.

MAKE THE APPOINTMENT

No matter how they respond to "what prompted you to call us in particular" you continue by saying "that's great. It's our tradition to invite first-time guests in for a complimentary consultation so that we can share some ideas back and forth and focus on creating the look that you really want for yourself. You're under no obligation but we'll be prepared to proceed at once if you're comfortable. Looking at the schedule, you could come in as early as this afternoon at 12:30 or 4:00 PM. Which works better for you?"

Notice that we're making the appointment without having revealed price. You'll discover that a large percentage of people will make their appointment at this point without asking the price question a second time. Fact is that when we get them in the door we may together decide on a complete makeover and who knows what level of spending that could involve.

However, if they ask the price question a second time it's then appropriate to reveal price. However, you only reveal price after first building value. The easiest way to build value is to describe step by step what the service includes. Once you build value on the basic service you reveal it's price and then immediately ask for the appointment again. For example,

> Our haircutting service includes a rich professional shampoo and reconditioning treatment. You'll receive a precision haircut. We'll take the time to finish off your look with all the gels and mousses and irons and dryers that are appropriate to create the full style so that when we're done you'll look and feel like a million. And, you'll learn how to maintain your look so that it looks fantastic until your next visit. And we'll be doing that entire service for you today for only $X, that's all, so would you prefer your consultation at 12:30 or 4:00 PM?"

THE ART OF PAUSING

Notice how we ask for the appointment immediately after revealing price and that we ask for it without even pausing for a second. This is crucial because your pause belongs after you ask for the appointment, not after you disclose the price. If you make the pause after the price a noticeable tension will often occur because the price has been made the focus. That tends to make everyone uncomfortable so the conversation could end without an appointment. Pause immediately after asking for the appointment. Once you pause you remain silent until the caller responds. During the pause turn your thoughts to a positive outcome to the conversation.

If the caller wants to ask another question be sure to respond effectively and always end again with the invitation to come in for a consultation. For example, if they ask "Do you have evening hours?" your response is to return to question mode and stay on top of the conversation. "Yes, we have evening hours. Is that more convenient for you? What time would work best for you? Great, then your consultation will be at 7:30 PM. Do you have a pencil? I'll give you parking instructions."

THE PATH OF LEAST RESISTANCE

You've built value and demonstrated professionalism. The consultation approach is the easiest way to get people to visit. Potential first-time visitors to any salon can naturally experience anxiety about what they're going to receive. With this approach, their fear of making a mistake is minimized because they're only agreeing to a no-obligation consultation. However, in reality, you'll discover that the vast majority will proceed at once with the service. Remember, they're calling because they want something done. This is not just some kind of a pass time!

TRANSFORM WALK-IN VISITORS INTO CLIENTS

Nine of ten people who walk into the salon should make a transaction before they leave. However, for a variety of reasons even a large percentage of walk-ins are lost because the interaction is mishandled. This is particularly disappointing when you consider that these folks are actually out there hunting for services and products. A stronger buying signal you couldn't ask for! Again, this represents an opportunity to improve our methods and build clientele and income with opportunity that's already coming our way.

INCREASE WALK-IN TRAFFIC

First of all, you want to encourage people to walk in. The more the better. It's true that some locations are naturally suited to attract walk-in business. However, a few easy things can be done to enhance walk-in numbers for any salon. It's a good idea to keep the salon door propped open to encourage people to enter. That one strategy alone can double your walk-in action.

Another proven idea is the sandwich board sign on the sidewalk or in the hall proclaiming "Complimentary Hair & Image Consultations Available Today" and an arrow with the words "Walk In." You want to have a sign like this placed so that foot travelers literally have to walk around it. Speaking of signage, I don't particularly care for "Walk-ins welcome." I much prefer "Appointments not always necessary. Please walk in." Naturally, if it's a nonappointment salon that's a different matter, but the "Appointments not always necessary" sign has more class.

ACKNOWLEDGE WALK-INS PROMPTLY

Have you ever walked into a store that appears totally vacant with no one available to help in sight? Chances are that if someone doesn't make an appearance to look after you promptly you'll simply leave. On the other hand, how about if you walk into a busy restaurant that already has a waiting line. You naturally, want to be promptly acknowledged to make sure you're on the list in order. You expect someone to be there to take down your name, give you an estimate of how long it will be, and even suggest you have a drink in the lounge.

Prompt acknowledgment is basic courtesy in any service establishment. We have a right to expect it when we're the consumer. Likewise, when we're the vendor we have an obligation to provide it. The failure to provide something as simple as prompt acknowledgment leads to lost revenue and disgruntled visitors. You can easily see how. So, make sure you recognize whoever walks in within a matter of seconds.

Obviously, it's counterproductive to spend any more than minimal time back in the staff room. I've seen situations where the salon is quiet and all the designers are in the back room talking about how slow things are while prospects are entering and exiting without even being noticed. How ironic! The big earners are always alert to walk-in opportunities and situate themselves to make an approach within seconds whenever someone appears. They spend their free moments lingering near the front of the salon.

Even if the salon is extremely busy, the walk-in must be acknowledged promptly. That may mean no more than holding up your index finger and nodding your head and mouthing "one minute please" and gesturing for them to take a seat until someone can greet them more formally.

GET OUT FROM BEHIND THE DESK

When someone walks in and asks how much does it cost, you must immediately realize that they're there to make a purchase. You want these walk ins. And you want to be able to convert most of them. Discussing prices when you're behind the barrier of a reception counter is not effective. When clients appear and start asking questions, move out from behind the desk and position yourself beside them. Just as on the telephone you want to ask them "Is the service for yourself?" "Have you been a visitor here before?" "What prompts you to visit us today?" Now you have got some information to work with; you've made the client the center of attention and you've started to build rapport.

The next step is to get the guest back into the salon area for a consultation. Say something like "Let's step back to the consultation are so we discuss some ideas to help create the style and image you really want. Here let me take your jacket and follow me." The goal is to get the prospective client into a consultation process somewhere other than the reception area. That way you'll have the individual's undivided attention and your likelihood of creating a transaction is dramatically improved.

POWERFUL CLIENT ATTRACTION STRATEGIES

GET THE MOST FROM INCENTIVES

Any marketing program, including a referral marketing program, gets more action when it includes an incentive. An invitation alone won't do half the work of an invitation combined with an incentive. The nature of the incentive you want to provide depends on how you want to position your salon and on your specific goals. Here are three popular types of incentives to understand and have in your marketing arsenal.

1. *Discount incentives.* It's a tried and true method. Research has shown time and again that dollars-off incentives always get more action than percentage-off incentives. Five dollars is a good number. More than that may cause suspicion. Less than that may not be motivation enough. However, you have to consider the price of your services. If you use discount offers make sure they are for broad-appeal services so that anyone can take advantage of the offer. If it's just for a root perm you've lost most of the market. The wash, cut, and style applies to everybody.

2. *Gift with purchase incentives.* A great incentive approach and a proven workhorse in the beauty business is the gift with purchase. When guests obtain something at regular price they receive something extra for free. You can use a lot of imagination in gift with purchase offers. You can combine retail with retail, service with retail, service with service or retail with service. You can have them receive the gift on the current transaction. You can have them receive the gift on a future transaction.

I myself like "gift with purchase" offers that provide up-selling opportunities, for example, a complimentary paraffin treatment with any chemical service. It gets those big spenders to sit down with the nail technician so they can learn about the beautiful benefits of sculpted nails. Or, how about a complimentary make-up lesson with every haircolor application? Now you have the opportunity to demonstrate and sell a make-up system. When you use the gift with purchase this way, you lead the guest into a sales presentation you'd like her to participate in anyway. And don't for a minute underestimate the power of a free dollars-off bounce-back gift voucher as a retention device.

3. *The free gift incentive.* Many salon professionals feel better about giving something away for free rather than using the discount approach. If you're positioning your salon in a more deluxe category, this makes sense. The strongest possible incentive you can give is something for free. Think of it as offering a free sample or a free demonstration or a loyalty reward. I know salons that will use the "complimentary consultation" or "complimentary color analysis" or even the highly potent "complimentary shampoo, cut and style" offer for targeted first-time visitors. Even "free ear piercing" can create fresh first-time traffic and leave room for up-selling. These offers can also be adjusted to build skin, nail, and body salon services with ease. The idea is to use the good will of the free offer to up-sell a myriad of unplanned service and product offerings. It works!

Remember, you can go to practically any department store and receive a complimentary mini-facial or make-up application and leave the counter with $50 or $100 worth of product easily.

MAGNETIZE FRESH CLIENTS WITH CROSS PROMOTIONS

When you stop and think about it, you'll be surprised how many businesses in your area are in a position to send clients your way. You'll want to harvest these referral sources and that means you must be proactive. The idea is to develop relationships with area businesses. You "cross-pollinate" with others by sending your customers to them in exchange for them sending their customers to you.

Whenever you approach other businesses you must do so professionally. Make an appointment, send a confirmation letter, show up on schedule, and be prepared with a specific proposal. Act like a professional and you'll get professional results. To get the relationship going you focus on what they're going to get from the transaction.

Noncompeting beauty providers can be a great source of referrals. If your salon doesn't provide the services they're offering then there's no conflict of interest. They're already dealing with people on how to maximize their appearance and there's no doubt many of their clients can benefit from your salon services.

Here's a short list of people you can link up with:

- Estheticians
- Make-up artists
- Tanning studios
- Hair only salons
- Mary Kay, etc.
- Massage
- Electrolysis
- Nail salons
- Beauty retailers

Let's say that you're approaching a tanning studio down the street. You start by mentioning how many of your clients ask you where they can go for tanning that's safe and clean. You indicate that you always believe in supporting area businesses and suggest that if your clients visited the tanning studio once that many of them are likely to become regular patrons. You then bring up the idea of getting some one-time passes that you can distribute to your customers to help get them in the door.

Now that they know what they're going to get out of the relationship, you continue by suggesting that it's only fair for you to offer something for them to give to their customers. So you tell them that you'll arrange a coupon that they can give away entitling their guests to come see you for a complimentary service. You might even mention that they can use it as an incentive for people who buy a package of tanning sessions.

APPROACH AREA BUSINESSES

Don't just limit yourself to other beauty providers. Here are some simple examples of businesses that you can develop a cross-promotional relationship with.

How about the area podiatrist? You make an appointment and mention that you have clients who need his care and ask if he would accept referrals from you. Naturally you tell him about your pedicure services and suggest that there are times when you'll be able to positively serve some of his patients. Have a team spirit. Make it a point to be the first one to send a referral to show your good intentions. Follow-up by telephone. Send a thank you note whenever you're sent someone.

How about the area jeweler? A new ring always looks better on a beautiful hand! When you introduce yourself mention that many of your clients have beautiful jewelry and that you'd like to encourage them to support the area businesses. Mention that if you could get them in the store to see all the wonderful items available that you're sure many of them would become customers. Perhaps the jeweler could provide some certificates for free jewelry cleaning. Then suggest that to help him sweeten the deal for people buying rings that you'd be happy to provide some gift certificates entitling purchasers to visit you for a complimentary mini-manicure or polish application. You're getting people who have money to spend and who care about their hand fashions. Perfect!

How about employment or temp agencies? They're in the business of helping secure employment for folks and everyone knows that the first thing a potential employer looks at is appearance. You can approach these agencies and position yourself as a salon that specializes in helping people package themselves for the job market. This is a tremendous benefit to the agency because there's only so much coaching they can give to a candidate about appearance without risking offense. However, the agency is certainly in a position to refer a prospect to you. The benefit to them of having access to a reliable professional service can be so great that you needn't offer anything beyond providing the reality of outstanding service in return.

Here are some additional businesses to consider:

- Weight loss centers
- Health clubs
- Fashion stores
- Bridal shops
- Photographers
- Department store/drug store beauty counters

These are just some of the obvious examples. Open your eyes and put your creative imagination to work and you'll be amazed at how many opportunities there are for business referrals and cross promotions right in your own community.

Summary

You'll want to have fresh new guests visiting all the time. That way you'll constantly improve demand for your services and experience meaningful income growth at the same time. When you have the confidence of a fully booked schedule all day every day your standard of living can only improve. Here's what we discovered:

- Family and friends are the first circle of contacts we have as clients—and we charge them!

- Cultivating client and colleague referrals is something you'll do for your entire career and you'll want to have an organized way to stimulate ongoing referral activity.

- The telephone is your most valuable tool for enticing fresh visitors, and improving your telephone skills will win you new clients immediately.

- Developing effective incentives and working with area businesses can be a highly effective way to cultivate first-time salon guests.

You must get them in the door before they will buy anything. Use the path of least resistance and focus on easy opportunities that already exist and you'll stimulate a steady stream of new business.

C H A P T E R 9

A Short Course in Ringing the Cash Register

"

You don't get paid for what you know, you get paid for what you do.

"

As salon professionals we only generate income when we sell services and products. That's right, I'm using the "S" word. We're salespeople. No matter what profession you pursue, the ones making the most money are the ones who aren't afraid to make sales!

WHAT YOU WILL DISCOVER IN THIS CHAPTER

- You'll come to grips with the impact that initiative has on your quality of life and standard of living.

- You'll discover how to provide more influential consultations that will stimulate impulse purchasing.

215

- You'll find out how to secure transactions will elegance and grace.

- You'll become more comfortable demonstrating and designing home maintenance systems that your clients will treasure.

THE TWO TYPES OF SALESPEOPLE

Now, there are two types of salespeople. Type one is the "order taker." Type two is the "order maker." Both the order taker and the order maker can have exactly the same knowledge and skills. Both can be perfectly delightful people. But in the end, the order maker generates more income and enjoys considerably more prosperity and satisfaction. Looking at this another way, the order takers are only working on the 30% of purchases that are planned. The order makers get those, plus access to the other 70% of purchases that are made on impulse.

My experience as a salon owner has taught me that the order maker retains more clients, generates more referrals, sells more home maintenance systems, earns far more on a per hour/per client basis, enjoys a more lavish lifestyle, and experiences higher self-esteem.

There are two main things the order makers do that the order takers do not:

1. They initiate.
2. They close.

THE POWER OF INITIATIVE

GET YOURSELF GOING

When you initiate, you begin. The first thing you have to get started is yourself. Your purpose, desires, and goals give you something to start for. What do you want from life? What role is your salon income and experience supposed to play in that?

Establish daily performance goals. It's basic to set financial performance goals. Annual, monthly, weekly, and daily goals are fundamental. Make it your first priority to set for yourself a daily financial goal. This is not just something to think about doing. This is something to have started days ago when you read about it earlier in this book.

Set a service goal and a retail goal. Write them down on your booking sheet so you can review them each time you check you schedule. Don't just add up what you have booked and hope for no cancellations! Force yourself to stretch. Make the goal challenging but attainable. If you have $100 booked then make your service goal $200. If you have $200 booked then make your service goal $400. Don't overlook the retail goal. If you haven't heard it before, your retail sales should be at least 25% of your service sales. If your service sales goal is $400 then make your retail goal $100.

Get psychologically focused for performance. Focus on the contribution you have an opportunity to make to your salon guests. Give yourself permission to play full out. Be a wild-eyed enthusiast. Be 100% positive. Confidently believe in yourself. Refuse to entertain feelings of doubt. Don't make justifications for failure but don't beat yourself up either. Remember that anything worth doing well is worth doing poorly at first! Focus on improving a little bit every day.

PLAN YOUR WORK

Next, plan specifically every day what you're going to do to exceed your financial goals. Your plan will involve:

- Selling add-on services to booked clients
- Selling retail products to booked clients
- Selling gift certificates and packages
- Selling a series of visits

- Effectively handling walk-in clients
- Effectively handling telephone inquiries
- Following up with recent visitors to meet additional needs

Let's focus on increasing the purchases of existing booked customers because this is the easiest area to improve fast. For each and every guest you want to have a specific plan of action to encourage impulse purchases of add-on services and retail products.

For example, stimulate add-on services. What's available at your salon? No matter what the range of service offerings, it's crucial you have a half-dozen or more extra things the client can have done and they'll potentially purchase on impulse. A nail technician for example, might come up with this list of service add-ons:

- You can offer two or three upgraded manicures.
- You can offer paraffin treatments.
- You can offer pedicures.
- You can offer to use special equipment.
- You can offer deluxe hand massages or cream treatments.
- You can "package" and offer upgraded levels of nail sculpting.

If you're working in a full service environment, the list is endless. You might want to bundle several services together and package these offerings with appealing names. Hit different price points so all clients can justify spending more in a way that fits their budgets.

FOCUS ON RESULTS

You might say "We already offer pedicure treatments . . . this isn't new." Yes, but let me ask you this . . . how are they offered? Is there a proactive effort with each and every guest to build desire and stimulate action, or are you merely there to take the order if someone asks for it? Therein lies the difference between poverty and prosperity—between order taking and order making.

To stimulate impulse purchases of services and products you must show initiative—you must put the idea on the agenda. Certainly, you can use the power of display and the printed word to introduce impulse purchase ideas. That's great as far as it goes. However, according to salon industry research of consumers seven of ten purchases are actually made because of the specific recommendation of the salon professional. So you have to talk. You must bring the ideas up in discussion before, during, and after your services.

This is another area where goal setting can be very productive. Set specific goals on how many "impulse services" you are going to sell in a given period. Be precise. "I am going to sell ten unplanned pedicures this week, and my goal for today is three. I am going to sell ten unplanned paraffin treatments this week, and my goal for today is three. I am going to sell twenty nail kit gift baskets this week, and my goal for today is five." Write it down where you can see it often. Set a new goal each day, week, month.

PROVIDE MORE INFLUENTIAL CONSULTATIONS

Make it a rule that every client gets a consultation before the service begins. In my mind this is the most important part of the visit. This is the time when impulse service purchases are stimulated. Don't just "wing it." Your consultation is a performance. Use charm and showmanship. Plan your show in advance. Know what you're going to introduce and when you're going to introduce it. Know what you're going to say, how you're going to say it, and when you're going to say it.

Be totally conscious of the fact that your interaction with each client is where the rubber meets the road for building wealth. Stay on top of it. Proper prior preparation prevents a poor performance. Have specific ideas about what would be beneficial, special, or just plain fun for the guest. I would recommend starting off the day with an analysis of who is coming in for an appointment and preparing a specific approach for each of them.

A very effective strategy in consultations with regular visitors is to let them know you were thinking about them before they came in. It's flattering and makes them feel even more special. "Jennifer, I was thinking about you before you came in today. And after you left the last time I began to think that the XYZ service would really make your nails stand out in a crowd. Let me tell you what it involves."

A word on first-time visitors and referred customers. If there's ever a time to get them involved in every service and product imaginable, it's on their first visit. Don't fall into the common trap of being timid when the situation calls for boldness! Attempt to consult them into a make-over quality of service.

You want to lead salon guests into a preplanned presentation or series of presentations. One effective way to do this is have them fill out a client card while they're in the reception area. You could include a couple of extra questions like "Are you looking for a change in style today?" and "Would you like your hair designer to suggest some style ideas that he feels would look good on you?" You will get almost all yes answers to these questions. That's full advance permission to make all the design presentations you want to.

With first-time visitors you'll want do all the things necessary to make a stunning and positive first impression when you introduce yourself. Let your guests know that the first item on the agenda is to share ideas back and forth to design an image and style that radiates their personality and good looks. Invite them to join you for a salon tour. Point out all the different departments and zones and areas to build interest and desire on the part of the visitor. Then direct them to where you perform your consultations.

Be interested, not interesting. The way to provide influential consultations is to be supremely interested in the needs and wants of your guests. They are their own most interesting subject. They are far more into the idea of us being impressed by them than being impressed by us. They are at the salon to look and feel better and to be the center of attention. The common mistake some designers make is to try to be impressive to their clients. Though it's necessary to build credibility, it's not a good idea to turn the focus away from them and onto ourselves.

ASKING THE QUESTIONS

Having the client as the sincere "center of the universe" while they're with us is going to create the environment necessary to allow for spontaneous and unplanned purchasing of additional services and products. Using the consultation method, the process revolves around two elements:

1. Having clients identify their own problems and articulate their own goals

2. Having the designer present solutions and motivate action

We get clients to identify their problems and share fashion goals by asking questions. In fact these questions are so fundamental to getting the consultation started properly that they need to be flawlessly prepared and consistently applied. Here are my three favorite questions for getting to the meat of the consultation:

1. "If you had a magic wand and could change anything about your hair today, what would it be?"

2. "What is the single biggest problem you encounter when styling your hair at home?"

3. "What problems did you have with your last haircut?"

Notice that these are open-ended questions and they also give you the opportunity to encourage the clients to elaborate on their responses by saying something like "what do you mean by that, exactly?"

222 • Chapter 9

It's always a good idea to reflect back your understanding of their problems or desires to make sure that they know that you understand exactly what they're talking about. "If I understand correctly, you're concerned about . . . , is that right?"

The "magic wand" question is my true favorite because it's so broad. When the salon guest responds by identifying a problem or expressing a desire you then say: "if I could suggest something here today that would solve that problem (or achieve that desire) would you like to give it a try?" It's the "If I could . . . would you?" questioning technique. And you'll discover quickly that it sets the stage beautifully for impulse purchasing.

When you ask that question you'll see that nine times out of ten you're going to get a yes or a maybe. What we're doing here is creating a mood of agreement before we get into the specifics of what we propose. That's valuable because you've been given full permission by an interested party to proceed with your presentation. Does that sound fair enough?

PRESENTING YOUR SERVICE SUGGESTIONS

I would recommend getting out pencil and paper and writing out a standard presentation for each and every add-on service offering you have. And don't fall victim to getting overly technical! Keep it simple and benefit oriented.

Because they've given you permission to proceed, you want to be specific in diagnosing the problem and prescribing the solutions. "Mrs. Guest, based on what you're telling me and my own observation and experience as a professional, you're suffering from X and need Y."

"X" is what the problem is. They're the same problems that come up over and over again. Have a precise name for each problem that customers encounter. Cuticle layer damage, moisture imbalance, protein loss, pattern baldness, color loss, lifeless hair, and lack of texture are the kinds of expressions you'll use to be specific about the problem that your solutions will solve.

Then have a specific name of each solution. For cuticle layer damage you could suggest your "intensive reconstructor treatment." For moisture imbalance it could be your "nuclear moisture bomb" and for protein loss you could prescribe your "deluxe keratin protein pak."

The same principle holds true for each chemical service solution you'd recommend. For your clients with soft, flat hair you'd call for "21st Century Hair Texturizing." For those experiencing dull, lifeless hair or pigment loss you'd insist on coloring services—whether it be the "Executive Blonding" or "Natural Color Glossing."

DESCRIBING THE SERVICE

The very best way to describe the service is to show them what it's going to look like when it's done. Your own before and after portfolio will boost confidence and help dramatize the results they'll receive. Other devices that you can bring into play can enhance your showmanship and get the client involved. Swatches, rods, brushes, foils, caps, and the like are selling tools as well as technical tools. It all helps build value.

Depending on the circumstances, you may elect to briefly discuss the steps of the service especially if you feel the guest needs a little reassurance. Use words that put the focus on dependable, predictable results. Build confidence by mentioning your use of the latest and most advanced and gentle formulations. Amplify your own expertise, experience, advanced education and artistry as a designer.

If your guest has a fear or concern, recognize that it's the same fear and concern that arise repeatedly. With perming, it's dryness, brittle hair, color problems, too much curl, perm fallout, and breakage. With coloring, it's another list of concerns. As you're describing the steps of the service, create the impression, for example, that rather than dry hair they'll find their hair is more luscious and luxurious after the service. Describe the results of the service by providing images of what they want and in that way you can dissolve potential fears.

MAKING YOUR PRESENTATION SPARKLE

Use client psychology to arouse desire. Keep in mind what motivates your clients to act on your suggestions. It has little to do with what they need. It has everything to do with what they want. And what they want is purely psychological. They want youth, beauty, freshness, health, glamour, fashion, confidence, security, freedom from worry, peace of mind, acceptance, prestige, and happiness. The style they have right now isn't getting the job done. They're dissatisfied. They want to feel better about themselves. They want to make a more favorable impression on others.

Emphasize that as a result of the add-on service, they'll receive the psychological payoff that they want. We discussed all this in great detail in the chapter on client psychology. Bring those points to bear in your consultations. The reassurance you provide will make your consultations irresistible and provide a real service to your client at the same time. Use passionate words. Talk to people where they live. Words paint pictures so go at it with gusto! We can talk to people intimately and please don't be afraid to do so. It actually bonds the relationship beautifully because you can share thoughts and feelings back and forth that wouldn't ordinarily be discussed.

Use testimonials. Relate your own experience or the experience of another client who had a similar situation. The structure for a testimonial is simple. You mention what it was like before, what service happened or what products were used to improve the situation, and finally, you share the happy results.

Janet, it looks like you have the same situation as another client, Rhonda. She was really beside herself when she came in. I know how awkward she felt in social situations. She had tried many drug store products that were purported to work but didn't help her at all. The home use program I recommended to her included products a, b, c, & d. She followed my instructions exactly. A few weeks later when she came in the situation had improved dramatically. She was in here just the other day and she's glowing with a confidence I've never seen. This kind of track record is exactly why I'm prescribing that you use these products.

TRIGGER URGENCY

People make impulse purchases for emotional reasons. However, they like to have a logical justification for their emotional purchases. When we trigger urgency we are providing a reason to act now. We're not going though the process of consultation just to provide information. We're doing it to secure a transaction at once.

An impending event in the client's life can help trigger urgency. A vacation, a business trip, a new romance, a reunion, a new job assignment, and other circumstances can help give a client logical justification for what they want. Don't hesitate to mention it when building desire for the transaction. "When you go out on that hot date with Bob tomorrow night you're going to mesmerize him with this stunning color." "When you walk into that high school reunion over the weekend heads are going to turn the minute you enter the room."

Fashion and style can also help trigger urgency, especially with clients who are socially conscious or want to be the first on their block. Exclusivity is one of the great motivators and you can trigger urgency by linking prestige, leadership, and recognition with having the very latest look. It will motivate many patrons to fast action.

Your promotions trigger urgency. Fear of loss is a great motivator. The promotion is only available for a limited time, so why miss out on the extra value? That triggers urgency.

But don't make the mistake of waiting for a promotion to begin earnestly building your income! In our time-sensitive and fast-paced world, the mere fact that the guest is in the salon and has scheduled this time to be with you is urgency enough!

Once you have made your irresistible suggestions, you must signal that it's time proceed with the service. This needs to be done with all the elegance and grace in the world. Here's how. It's really simple to do (Fig. 9-1). To conclude a service add-

FIGURE 9-1 You conclude your presentation by asking for permission to proceed, "Does that sound fair enough?"

on say, "The schedule is arranged for you so we can proceed on this at once. Does that sound fair enough?"

You must ask the question! I call it the magic phrase "Does that sound fair enough?" It appeals to the patron's sense of fair play. There's nothing at all offensive or discourteous about it. You're simply alerting them that the opportunity exists to proceed and then giving them the opportunity to take advantage of the opening. "The schedule is arranged so we can go ahead and proceed on this at once. Does that sound fair enough?"

Nothing will happen if you don't ask the question. You'll simply wind up being an information booth and not an order maker. Research indicates that 85% of presentations are made without any request for the order. And when there's no request for the order, there's no sale. In fact a tremendous amount of business is left "on the table" because the cosmetologist didn't ask for the opportunity to proceed. That's a lot of money lost!

We live in a culture where people expect to be asked. As a matter of fact, some of your customers feel downright uncomfortable about asking for something extra. They're concerned that they may be disrupting your schedule. They may fear appearing demanding or rude. People in our culture feel it's more polite to wait to be asked. We were raised that way, for heaven's sake! Don't ask . . . wait until you're offered!

I remember not that long ago when I was out of town and treated myself to a marvelous manicure. The nail technician made it a point to let me know about all the products that were being used and how they can be used at home with great results. Then when we were at the cash register and she was writing up the service ticket not another mention was made of the products. What a mistake!. Apparently she was waiting for me to bring the idea up. Well, like it or not, that's not how it works.

When you signal the opportunity to proceed and say "Does that sound fair enough?" you're going to be pleasantly surprised at the number of people who say "Yes, that sounds fair enough to me."

Likewise, some are going to say "No, not now." If it's a matter of scheduling or budget you can always proceed by arranging to perform the service in the immediate future. Say something like "It would be a shame to pass up on this opportunity while it's fresh in your mind. How about if I go ahead and arrange for it later this week or next week. Which would work better for you?" Then say "I'll go ahead and pencil it into the schedule and we'll proceed then, fair enough?"

If it's an outright no, turn it into a positive by trying to figure out what you can say or do differently the next time to improve your closing percentage. And, don't beat yourself up. Rather, pat yourself on the back with the knowledge that you're going through the pain necessary to achieve the gain in income and quality of life that you're aiming for. You're doing what it takes to double your income. And if you keep on doing it your effectiveness and self-esteem will blossom. See for yourself!

CHARMING RETAILING

Part of the service that we provide involves demonstrating and designing a home maintenance system for each salon guest. It's part of how we distinguish our excellence and provide a long psychological afterglow to the patron's visit.

INTRODUCING THE HOME MAINTENANCE SERVICE

Before guests even walk in the salon, they're receiving messages about our products. Window dressing and decals plus the display cases containing the product itself all put retail on the agenda before the visit is even underway. Certainly, by the time guests are in the reception area they may be exposed to "testers" or product brochures or other information. They'll encounter product displays more directly.

By the time the guest is involved in consultation, we're actively discussing home maintenance, especially if we're getting into chemical services or if a hair or skin or nail problem is uncovered that needs ongoing attention. It's all very low key at this point. "Janice, before you leave I'm going to get you a special shampoo and conditioner that will add life and luster to your haircolor so that it will stay fresh longer."

PRODUCT DEMONSTRATION

At the back bar we definitely want to be discussing what products we're using. This is an area of neglect that you'll want to address at once. I've discovered that in the vast majority of instances the patron is not given any information whatsoever on what products are being used at the back bar. This is a serious error. The back bar provides us the very best opportunity to discuss and demonstrate the products we're using and to plant the seed for their eventual purchase.

A person can go into any department store, walk up to a skin care or make-up counter, and receive a mini-facial complete with detailed information and demonstration on every item that being used—and receive this for free. Yet when we have them in the salon where they're paying for the information, and we neglect to provide it, something's not right It's unbelievable!

As a matter of practice you'll want to clearly state exactly what you're using by brand and by specific product name. Say why you're using it. Say how you're using it. And discuss the fact that they should be using it at home.

Remember to use the word "you" when discussing product benefits. People aren't that interested in "it."

> You use spray and gel so you naturally get a lot of build-up on your hair, Mrs. Jones. So your shampoo will be with XYZ brand deep cleansing shampoo. You'll notice that your hair and scalp will be squeaky clean by the time your shampoo is done. And you'll find the effervescent aroma invigorating. As a matter of fact, Mrs. Jones, you'll find that a bottle of this will be nice to have at home. You'll want to use it before your deep conditioning treatments for sure.

Every time you bring a product into play you discuss specifically what you're using, why you're using it, how you're using it and the fact that the client should be using it at home. When you get into fixatives and sprays at the styling station demonstrate the product. Talk about it. Discuss how it's used.

Get the client involved. You've heard before that it's a good idea to put the container in the client's hand? This is the time to do it. It builds automatic curiosity and interest. Put the blow dryer in her hand. Make sure she can duplicate at home what you're doing in the salon.

THE FOUR LAWS OF RETAILING

When the cape is removed and the client gets up, our next move is to escort her personally to the home maintenance system display area—otherwise known as the retail area. And we keep in mind the four laws of retailing.

Law I: The Law of Consistency. We offer the retail service to each and every guest. We don't prejudge. We don't assume anything. We offer it automatically as part of our service. Even if we're extremely busy, we still take a few minutes to design the home maintenance system.

Law II: The Law of One Line. When we are designing the system we focus on products from a particular line. People just automatically believe that they're going to receive the best results when they use a series of products from a single manufacturer. Consumers just accept the idea that products from one line are chemically compatible and that results will be sacrificed with mixing and matching.

How would you react if you went into the skin care section of your favorite department store and were told you should buy your cleanser from Esteé Lauder and your toner from Clarins and your scrub from Clinique and your day cream from Arpel? You wouldn't accept it. The same principle holds true in the salon.

I understand some clients have tried a variety of brands and have their favorites from each. If they know what they want, provide it. However, when you're designing the system, abide by the law of one line.

Law III: The Law of Large Sizes. People automatically believe that when they purchase the larger size they receive a better value—a lower per ounce cost. It's generally true. However, I've often seen situations where Mrs. Johnson will indicate that she would like to try some of that shampoo—and then the stylist goes and gets the travel size or whatever the smallest size available is! I've seen some go hunting for a sample to give away!

On our daily use products, the biggest seller should be our 12-oz or 16-oz sizes not our 6-oz or 8-oz sizes. The larger sizes are the ones people prefer. And encourage them to move up to the family size, especially for their daily use shampoos and conditioners. That way they'll receive the most value and you'll effectively eliminate the risk of them shopping for shampoo at the drug store between visits. Also, when consumers purchase the larger size they're much more apt to leave it in the shower or bath so other members of the family can use it. That way you get your product on more heads in the house, which will increase the amount of product used and the frequency of purchase. Once consumers have stepped up to the family size you'll discover that they wont go back to the small size again. So effectively, you'll begin to see an increase in the average value of each retail transaction.

Law IV: The Law of Large Numbers. We design an entire system for the client. We don't just stop with the daily use shampoo or daily use conditioner. Rather we recommend and integrate the entire array of products that they want to have at home to use. It could easily come to seven or eight items.

- Daily use shampoo and conditioner
- Deep cleansing shampoo and deep conditioner
- Some sort of a mousse or styling lotion or gel
- Hair spray and some sort of a glossing or high-shine finishing product
- Sometimes a scalp lotion or a food supplement or other product

We need to let them know about all the items and how they work together. Actually, we're summarizing the information and demonstration we provided during their visit. All the products that we used and talked about become part of their home maintenance system.

People love this level of service. When you've gone to the skin care counter at the department store don't they tell you about the cleanser, toner, scrub, masque, day cream, eye cream, night lotion, etc.? They give you all the information. How do you feel about it? Do you take offense or are you interested and fascinated? If you had the ready cash on hand wouldn't you want to take pretty much everything that was offered?

It's the same thing in the salon. Grooming products are fun. We discussed the psychology of grooming and how clients make a positive statement to themselves about how they value themselves when they use our products. And it's the energy that you infuse into the product by the benefits and values you link to the use of the product. That's how you automatically infuse your energy right into the bottle. And that energy is carried through the medium of the product and creates a long, positive, psychological afterglow for the client.

CONCLUDING THE TRANSACTION EFFECTIVELY

Each product in the system is put on the reception counter as it's being described. As you're putting the last item on the counter make a summary statement: "Using this system the way that we've talked about will give you the magic potions to have a great hair day every day. Go ahead and give them a try to see for yourself how well they'll work for you. Here let me get a bag for all this." If there's a receptionist behind the desk you might say "Janice, would you get a bag for all this?"

Now what we're doing here is presuming the transaction. Don't skip a beat. Don't show any hesitation. Move forward smoothly and gracefully. If the client doesn't want the entire system she'll let you know. But you're going to be surprised how many will take the entire system. It's a big priority for people. Even people of modest means will often make substantial purchases of products at the salon. It's not up to us to second guess or decide how much the other person can afford to spend. That's the individual's business. Our job is to provide the best service we can.

If the client hesitates about obtaining the entire system, we can always whittle it down in the sizes offered and the overall number of items until we arrive at a selection that meets their needs. But by starting with the full system we'll end up doing a lot better than if we had started with a single tube of shampoo.

As we pare the system down, we explain why some items are essential and why some can wait. Arrive at a selection of three or four items. Conclude with a summary sentence: "This grouping will get you started off real well. Does that sound fair enough?" When you ask does that sound fair enough you'll want to look the patron in the eye with warmth, nod your head ever so slightly, repeat the word "yes, yes, yes" to yourself silently, and then remain totally quiet until the guest responds. You'll surprise yourself at how often they'll say yes. That's how you bring that retail transaction to a sweet conclusion.

A final strategy is to have a few goodies that you can throw into the deal. Perhaps some samples or a foil pak or a small product premium. These are what we call deal sweeteners. "Mrs. Johnson, since this is the first time you've had your hair colored here I'll tell you what I'll do. You go ahead and take this system of home maintenance products—and this little bottle of hair gloss—the one that gives your hair that real radiant shine—I'll go ahead and put it in the bag along with the other items as a free gift and a special welcome to our salon. Does that sound fair enough?"

This approach is very powerful. Folks will jump at it. It's another example of dropping a nickel to pick up a dollar. Your increase in retail transactions and your improved good will with the client will end up paying you back substantial dividends.

SUMMARY

Opportunity sits in your chair with every new patron. Folks are looking for something new and exciting. Motivate them to action and they'll reward you at the cash register.

Here's what we learned:

- A specific plan of action for stimulating impulse purchases with each guest is essential.

- Uncovering client problems and confidently suggesting solutions is the cornerstone of providing an effective consultation.

- Patrons will wait for us to close the sale and it's part of the role we play as cosmetologists.

- Home maintenance systems are an important part of the service, and retail transactions can be concluded with consistency and charm.

We can know all the ideas and be no further ahead. The key is action. The opportunity is real and the time is now. And the transformational power is profound. We will be rewarded not for what we know, but for what we do!

Bibliography

Amen, Daniel. *Don't Shoot Yourself in the Foot: A Program to End Self-Defeating Behavior Forever*. New York: Warner, 1992.

Branden, Nathaniel. *How to Raise Your Self-Esteem*. Toronto: Bantam, 1987.

Braude, Jacob M. *Complete Speaker's and Toastmaster's Library*. 8-Volume Set. Englewood Cliffs: New Jersey: Prentice-Hall, 1994.

Covey, Stephen R. *The 7 Habits of Highly Effective People: Powerful Lessons in Personal Change*. New York: Fireside, 1989.

Dudley, George W. and Shannon I. Goodson. *The Psychology of Call Reluctance: How to Overcome the Fear of Self-Promotion*. Dallas, Texas: Behavioral Science Research Press, 1986.

Foley, Mark D. *The Best of Mark D. Foley: Supernatural Salon Success Strategies for all Beauty Professionals*. Calgary, Alberta: Mark D. Foley, 1995.

———. *Supernatural Salon Phone Skills: The Insider's Guide to High Powered Salon Phone Procedures*. Calgary, Alberta: Mark D. Foley, 1995.

———. *How to Achieve Supernatural Salon Income*. 3-Video Set. Calgary, Alberta: Mark D. Foley, 1994.

Givens, Charles J. *Super Self: Doubling Your Personal Effectiveness*. New York: Simon & Schuster, 1993.

McGee-Cooper, Ann and Duane Trammell. *Time Management for Unmanageable People*. New York: Bantam, 1994.

Moine, Donald J. and John H. Herd. *Modern Persuasion Strategies: The Hidden Advantage in Selling*. Englewood Cliffs: New Jersey: Prentice-Hall, 1984.

Presnall, Lewis F. *The Search for Serenity and How to Achieve It*. Salt Lake City, Utah: U.A.F., 1959.

Robbins, Anthony. *Unlimited Power: The Way to Peak Personal Achievement*. New York: Ballantine, 1986.

Rohn, Jim. *The Treasury of Quotes*. Dallas, Texas: Jim Rohn International, 1994.

Index

Note: Page numbers in **bold type** reference non-text material.